LEARNING THE BIRDS

LEARNING THE BIRDS

A Midlife Adventure

SUSAN FOX ROGERS

THREE HILLS

AN IMPRINT OF

CORNELL UNIVERSITY PRESS

ITHACA AND LONDON

Some of the chapters in this book have appeared in previous forms:

"Dawn Chorus" appeared as "Dawn Chorus: Reliving FDR's Wartime Birding at Thompson Pond," in *The Hudson River Valley Review: A Regional Review*, Fall 2020.

"The Other Leopold," first appeared in *Michigan Quarterly Review*, Summer 2019, vol. 58, no. 3, and was republished in *Best American Essays, 2020*, edited by Robert Atwan and guest edited by Andre Aciman, Houghton Mifflin, Fall 2020.

Photographs by Christina Baal (in The Other Leopold) and Peter Schoenberger (in I Wish I Knew; Snow Bunting; Learning the Birds; Bicky; Methinks; Twitching; Christmas Bird Count; Don't Move; No Other Everglades; Little Brother Henslow; Good Bird; Guided; Chiuit; A Perfect Fall Day; Surviving the Winter; Rusty Blackbird) are used by permission of the photographers. All other photographs are by the author.

First published 2022 by Cornell University Press

Printed in the United States of America

Library of Congress Cataloging-in-Publication Data

Names: Rogers, Susan Fox, author.
Title: Learning the birds : a midlife adventure / Susan Fox Rogers.
Description: Ithaca [New York] : Three Hills, an imprint of Cornell University Press, 2022. | Includes bibliographical references.
Identifiers: LCCN 2021042165 (print) | LCCN 2021042166 (ebook) | ISBN 9781501762246 (hardcover) | ISBN 9781501762253 (pdf) | ISBN 9781501762260 (epub)
Subjects: LCSH: Bird watching. | Birds. | Natural history. | LCGFT: Essays.
Classification: LCC QL677.5 .R64 2022 (print) | LCC QL677.5 (ebook) | DDC 598.072/34—dc23
LC record available at https://lccn.loc.gov/2021042165
LC ebook record available at https://lccn.loc.gov/2021042166

For Becky

You have learnt from the birds and continue to learn.

ARISTOPHANES

CONTENTS

I Wish I Knew   1

Snow Bunting   13

Learning the Birds   23

Bicky   37

Methinks   47

Florence   61

Twitching   81

Christmas Bird Count   99

Don't Move   117

No Other Everglades   131

Little Brother Henslow   143

Interlude: The Other Leopold   155

Good Bird   167

Guided   181

Chiuit   197

A Perfect Fall Day   209

Surviving the Winter   215

#1 Birder   227

Rusty Blackbird   241

Dawn Chorus   253

Little Blue   265

So Much to Learn   273

Acknowledgments   279

Notes, Notes on Notes, and Further Musings   283

LEARNING THE BIRDS

I Wish I Knew

Las Piedras River, Amazon, Peru

For a moment the howler monkeys stopped howling. Without the boom of their treetop calls it was as if the earth had stopped breathing. I followed, holding my breath as I sank into my dinky pack raft, my camera cradled between my legs, binoculars strapped to my chest. The raft felt like it might be losing air, slowly deflating to lower me into the muddy Las Piedras River in the Peruvian Amazon. And then the chorus resumed: a Screaming Piha screaming, a flock of Mealy Parrots (so green and more fun than their names imply) bulleting across the lightening sky hollering their intent, and Scarlet Macaws soaring high overhead, their long tails trailing as if off to market, calling out the chaos of life. A half dozen chips and songs I did not recognize, tantalizing calls and songs from the green green world, joined the chorus.

The boat hugged me like a sagging bassinet as the current pushed me along, downstream to the research station where I had begun my journey at 4:30 in the morning. I peered ahead to the next bend to see if my three companions were waiting for me. But they had vanished, leaving me alone in the jungle.

But not really alone. Yellow-billed Terns coursed upstream. Amazon Kingfisher cackled away like kingfishers do. Hoatzin ruffled about in the streamside bushes. A capybara family swam the far shore in a single file. And, always, the howler monkeys were booming from the tops of trees.

Into this ecstatic musical medley, I heard the puttering engine of a boat. No doubt it was one of the long, narrow, wooden boats taking Brazil nuts into Puerto Maldonado to be sold. The wake of the boat would surely upset my precarious raft, so I made toward shore, to secure myself to a branch or log. As I approached the shore, the log I aimed toward began to move, turned, showed its intent little eyes, and slid into the water. The caiman vanished into the turbid river. I laughed, half nervous, as I realized that for all I heard and saw, there were infinitely more creatures out there, going about their secretive lives. This was paradise.

An hour into my float, and still three hours to the station, a long-legged shore-birdy bird, with hints of blue and green on its back and a great red eye, appeared on a sand bar. I nudged my boat onto shore and slipped out of the raft, camera in hand. As I crouched to watch the bird, it approached, not skittish about having its portrait taken. In that moment, my sense of discovery was acute. Though I had spent time with the *Birds of Peru* guidebook, I could not name this bird. I felt a rush of excitement finding this beautiful unknown-to-me bird.

This is why I bird. The thrill of quiet adventure. The constant hope of discovery. The reminder that the world is filled with wonder. When I bird, life is bigger, more vibrant. And the Amazon, not surprisingly, is the perfect place to nourish such a sense. Though in that moment I was relishing feeling overwhelmed by the natural world, I don't need excess, abundance to get my adrenaline flowing, to have my curiosity piqued. All it takes is an encounter, close or unexpected, something small, even subtle, whether traveling, or near my home in the Hudson Valley of New York to make me thrill to the day.

The boat, the river—this is how I have spent my life: in the outdoors, on adventures far and near. I have hiked, back-country skied, kayaked, and rock climbed around the country and world. But the bird that held my attention tiptoeing the shoreline, this sort of birdy encounter was new in my life. I came late to my love of birds. Though on those hikes and paddles I wanted to know the birds, my efforts to learn were halfhearted, remained a wistful "I wish I knew." But once I realized that birds gave texture, meaning, excitement to the everyday, I gave over, converted to the tribe of binocular-toting people in hats and practical pants.

My conversion story begins on a spring evening in 2009. It was the end of the semester where I teach at Bard College. I was sitting in a cabin tucked in the woods near the Tivoli Bays in New York's Hudson River Valley with a few of my students, who were drinking beer and chatting with the exhaustion and euphoria of those who are about to graduate. While I looked forward to a summer writing and kayaking the Hudson River, they were all wondering what to do with the rest of their lives.

I remembered being twenty-two, graduating with my philosophy degree from Penn State and thinking as they were: What now? I wanted to travel, explore, rock climb, not get a serious job. So I left for France, my mother's home country, with a thousand dollars I had earned bussing tables, and spent the next year and a half working the grape harvest in Champagne and as ground crew for a hot-air balloon company. In between jobs, I wrote in my journal, rock climbed on the boulders at Fontainebleau and the seaside cliffs of the Calanques. Like a cat, I knew I'd land on my feet. And when I didn't, my sister Becky caught me. She was living in Paris, deep into graduate school, so not rich but always generous. She often fed me, let me sleep on the couch, and found me work as a carpenter's helper where I spent hours sanding overhead beams, sawdust sprinkling into my eyes and ears. Though during those years in France I had been anxious about money, had ached with loneliness, and had tossed through many nights wrestling with the heavy question of what I was going to do with my life, what I remember most is the sense of freedom I had, the privileged ability to say yes to life. I had had it easy, while these graduating students needed to find real work, were wondering about paying off college loans: 1983 and 2009 seemed worlds apart.

The house cat laced its way between my legs, and I rubbed its ears. And then a "spiral of white gold" poured out of the hardwood forest, entered through an open window, and seized the room like a tentacle of sound. Hollow and holy, what I heard resonated so full and complex; it had texture like a silk accordion. Silent, the students looked at each other and then at me, Ali's eyes wide, Sam's eyebrows arched. Was this creature going to come through the window and join us? The sound emerged again, winding us closer together.

"What *is* that?" asked Ali, a young woman who grew up in the Hudson Valley. Perhaps she was asking if it was a bird or some other animal, or a visitation from another planet.

I was supposed to know what creature was making this sound, and not only that, have words to describe it. I taught a nature writing class, asked students to go out, pay attention, then write about what they saw, smelled, and heard. I should have been able to name a creature with a song so distinctive, a creature that shared woods familiar to me. That I did not know this neighbor made me feel like a fraud.

"I wish I knew," I said.

Humiliation rests at the heart of my conversion story.

The next week, Ali brought me a card. Her words of thanks for being a teacher have stayed with me. (Not surprisingly, she too has become a teacher.) With the card, she included another gift, a book titled *The Backyard Birdsong Guide*. A number next to each image of a bird led me to the corresponding recording of the bird's song. With a push of the button on the attached plastic audio box, the song emerged. I was uncomfortably aware that the book had a beginners feel to it, the sort of book that had the potential to gather dust on a shelf. But it was just what I needed. I went through each bird, beginning with the Canada Goose and moving on to the perching birds, pushing the audio button to hear songs exotic (American Woodcock) and familiar (Barred Owl), in a rush to find the one that would match what we had heard sitting in the cabin. I finally arrived, just after the Wood Thrush and just before the Gray Catbird, at the Veery.

I looked at the reddish-brown thrush with its speckled throat and pot belly. Such an ordinary looking bird making such an extraordinary sound. Again I tapped the button to listen to its song, and what filled my little living room in Tivoli was a sound described as "the chiming of bells or the gentle sobs of organs." Organs in a church, song reverberating against stone walls, centuries of suffering and love. And in that moment thirty-five years of *I wish I knew* converted to *I will learn the birds*.

It's a tidy story, and in this way follows in the footsteps of many birders, who can tell what bird set them on their birding path. One of my favorite conversion stories is told by Frank Chapman, longtime curator of birds at the Museum of Natural History, editor of *Bird-Lore*, and founder of the Christmas Bird Count, now one of the great annual birding traditions. He wrote in his autobiography about the moment he decided to focus his life on the birds. It was the end of the nineteenth century when Chapman found himself working in a bank in order to make a proper living. He met a man, also a banker, and what he saw he did not want to be or become. "As I looked at him there suddenly sprang into my mind with the force of a revelation a determination to devote my life to the study of birds. . . . The sudden and convincing manner in which it was formed had in it something mystical which seemed to take the matter wholly out of my hands."

Chapman was surely aware that he was writing about his commitment to the birds as if it were a religious experience. His decision is fated, driven by an outside force: God or the birds themselves. Sudden, soul-moving life choices—no one did it more dramatically than St. Augustine, who chronicled his path in *The Confessions*. And even though his conversion occurred in 386 AD, it remains, despite the fact we do not live in an age of miracles, a good model.

Augustine's conversion did not come without years of hesitations, of wanting his salvation but not being able to give over: "Give me chastity but not yet!" he cries in what is perhaps my favorite line in the book. Chapman, too, spent years birding before he gave over, left the bank, and became a central figure in the history of American bird life. I, too, spent years wanting to know the birds, always curious about a raptor spied spinning in the sky or the songs that accompanied me when I hiked in the Catskills. But as the years marched on I kept hesitating: "Give me birds, but not yet!"

Nineteenth-century author and naturalist John Burroughs writes in his essay "The Art of Seeing Things" of meeting a woman who claims that she wants to see and hear birds: "No," he responds. "You only *want to want* to see and hear them." For years I was that woman, thinking that eventually I'd get to the birds, perhaps realizing that I could not learn by accident or in the cracks of life. I would have to give over completely.

To see birds, Burroughs clarified that "you must have the bird in your heart before you can find it in the bush." You have to *want* them with a particular birdy desire. This is why Chapman's story is not unusual in the bird world: birders rarely dabble. We are driven by passion; we are a tribe of believers. Since there are so many of these tales of the light turning on and illuminating the true path of birds, they have a name: the spark story.

My spark story is that Veery's song, the moment when I knew I had to learn the songs of the birds, wanted to be familiar with the everyday and the rare, wanted to be able to walk down a path and point to a buzz in a tree and say, "Blue-winged Warbler." And yet I also know that just one song can't be the only element that played into my conversion.

That Veery sang just when I needed the birds. I had a career, a calling in teaching. But like many teachers, I had thrown myself in fully, loving and mentoring my students as if they might be my own kids. Now I yearned to be the student not the teacher, to be the one learning, the one mentored.

At the time I embarked on my bird journey I was finishing my first book—the hardest thing I had ever done, I said to anyone who would listen. Writing was isolating, not just from friends and family but from my outdoor pursuits. I spent long hours at a desk, indoors, in silence, writing about being in a kayak on the Hudson River. At times this seemed absurd. By the time I handed the book off to my editor, I was eager to reconnect with the world, wanted to know the world in greater depth, to experience a greater texture.

Above all, when that Veery sang, I was taking stock of my life. What I found was that I was tired, not of the world, but of myself. I wanted to take more risks, live bigger. I didn't want to think that who I had become was who I would always be. I value steadiness, but I crave change.

What I had to work with was sobering. I was single with no children, and though I was grateful I wasn't adding to the overpopulation of the planet, I sensed I had missed out on one of life's great journeys. Both of my parents had recently died, and my one sister lived an ocean away in Paris. As for romantic relationships—I had made enough bad choices that it seemed wisest to retire from love. But understanding the shape of my life—that there was not going to be a happily-ever-after story, that I would not write a best-selling book, that I would never be rich, or even the simple fact that my legs would always be too short—was also liberating: I could abandon certain ambitions and follow what made me happy. The birds, I realized right away, made me happy.

I did not grow up with robust parents strapping on my snowshoes, lead-ing me on sylvan outings, and feeding me berries from a midwoods bush. My father loved words, books, often reading through the night. His par-ents had worried about their tall, skinny son, so devoted to daydreaming. My mother loved her daughters and her home country, France, which she and her family left in 1941. Yet if my parents did not take me to the woods, the woods were never far from where I grew up in central Pennsylvania.

It was the mother of my childhood friend Sonia, Mrs. Caton, who en-ergetically took her daughters as well as my sister Becky and me into the woods on a weekend. We built forts, dammed little streams, and picnicked in what felt like wilderness. This only made me want more. I begged to sleep outside under the tall pine trees next to our house in the middle of State College, the town that cradles Penn State. Becky and I prepared as if for a major expedition to the front yard and eventually, sometime past midnight, tucked into our cotton sleeping bags. Those bags remain vivid in my memory: the outside a dull green, the inside off-white with images of hunters, long rifles at their side, accompanied by hunting dogs. How fond I was of that first square-shaped, soft sleeping bag. The light from the corner streetlamp, the prickle of pine needles (ensolite pads were not yet, but would soon become, a part of my gear), and the image of the boogey-man would keep me awake through most of those nights. The edge of fear brought on a queer, protracted rapture; I needed to be outside.

From those first tame tastes of the outdoors, I lived for adventures. In my teens I joined an Explorer Post sponsored by the local outdoor shop. I slithered into caves, carbide lamp lighting the way, sewed my own down jacket, learned to wax my wooden cross-country skis, and, at age fifteen, tied into the end of a rope and scaled my first cliff. I spent the next years, through my teens, twenties, and early thirties, dedicated to rock climbing mostly at the Gunks, a band of cliffs outside of New Paltz, New York, but also on the steep cliffs of Eldorado in Colorado and on the domes of Tuolomne in California.

That I turned to birds may not seem a big step for someone who has spent her life outdoors. But rock climbing in the 1970s was not a nature-loving pursuit. It was irreverent, rebellious. If one image can capture the moment, it is Dick Williams climbing over a large overhang at the Gunks naked. Dick was one in a group that called themselves

Vulgarians—counterculture, rule breaking, and happy to pee off the top of a cliff after doing a climb named Dick's Prick. Most climbers of that era did not know their trees and ferns; they were rebels, self-proclaimed dirt bags. In my mind nature-loving people were soft, nerdy; bird lovers were dowdy or cranky old ladies, or men who wore pants hitched up at the waist. That did not appeal to someone who wanted to climb hard and who, like many good girls from good families, was awash in the romance of the bad boy.

Of course there were always birds when I climbed—the Turkey Vultures spiraling over the valley and the Wood Thrush that spun their songs at dusk as we walked out the carriage road at the Gunks, on the Mohonk Preserve. But the birds were background music to the buzz of adrenaline after a day of pulling my way up cracks and stepping out overhangs. If I was curious to know what was singing or flying, I didn't take the time to figure it out. Looking back, I'm now embarrassed at how many years I devoted to listening to my own heartbeat. Through those years, it was as if I was having a one-way conversation with nature that was all about me.

Age forty-nine, 2009, I finally was ready for a real conversation, one where I did most of the listening. I knew that the birds had something to tell me and wondered what I might learn. Perhaps learning and what I learned might give me what to do with the rest of my life.

The first three years after I walked into the world of birds took me from my home in the Hudson Valley of New York to Florida, Arizona, and Alaska. My conversion to the birds was physical, an outer journey. My goal was to get to know the birds and the world that surrounds them. I soon found that this bird world is one with its own logic, rules, and ethics, a place with its own language and literature, its own art and history, its heroes and a few heroines. I felt like an explorer who had landed on a distant continent, or more rightly a pilgrim who had arrived, after years of walking, at Compostela.

My journey involved that daily walking but also a lot of reading; I reveled in the volumes written about the bird experience. My reading allowed me to look over my shoulder for insight from John Burroughs as I trekked up Slide Mountain looking for a Bicknell's Thrush, or to Florence Merriam Bailey as I first met a Clark's Nutcracker. Knowing the birds through the lens of history, whether that person I followed through woods and lane was the president of the United States or a notorious murderer, added texture to the experience. And because the women seemed not-surprisingly but sadly absent in the world of recorded avian observation (though not in the actual world of birding) I looked for them, finding my favorites, especially Florence Merriam Bailey and Olive Thorne Miller. Reading their words and about their lives gave me new ways to think about birds. New ways to think about life.

And the birds inspired me to write. I am an essayist, a person who writes essays. I write to figure out what I think and come to understand how I feel. Through writing I see the world, and I see my own life. The essay form suits me: it is imbued with a spirit of exploration, and at the heart of most essays rests a journey whether physical, spiritual, or emotional. My story with the birds is all three.

Essays are not neat, leaving much unknown, with often incomplete ideas. This is not unlike birding: glimpsing a bird without fully being able to see and identify it. The unfinished-ness of the essay, the unknowns of birding, this appeals to me because life is messy, complex. There are a lot of moving parts to a life, to a bird. Trying to learn the birds was a great and surely impossible goal. And so off I went, meeting and naming, learning the birds.

Still an hour float from the research station, I once again heard a boat moving toward me. I saw Paul, our jungle-man guide, who allowed me great freedom on this trip, running the engine. In the long wooden boat sat his mother at the bow, the wind in her black hair. *How sweet that he's taking his mother for a morning ride*, I thought.

They slowed as they approached me, the boat making a wide arc to turn around.

"You good?" Paul called across the water.

I gave a thumb's up. I smiled wide as he headed back downstream. So he had come out to check on me, perhaps imagining I had gotten lost in the jungle or lost in jungle time.

"What took you so long?" he teased when, elated, I finally wandered up the hill to the station, to scrounge up a late breakfast. He had imagined my float would take an hour and a half; the other boaters in my party had been back for some time already.

"The question is why I ever came back. Let's start with the Sand Night-hawks roosting in the middle of the river. And there's a shorebird I met." I picked up my fat guide to the birds of Peru and flipped through to find the bird that had so captivated me on the shores of the Las Piedras River. "Southern Lapwing," I said. "You can't imagine how beautiful it is."

Snow Bunting

Ashokan Reservoir and East Kingston, New York

Near dawn, Peter and I walked along the mud and rock edges of the Ashokan Reservoir looking for a Snow Bunting. The wide-open space, the
rocky land underfoot, the short tawny grasses that sprung from the ground
all led me to imagine that this was the tundra, not the Hudson Valley.
I squinted as I elaborated on my tundra fantasy: we were alone, dropped
off with our tent and sleeping bags, just us and the Snow Bunting in a
spare, frosted world.

It would have been easy to give over to this fantasy except that in the
sharpness of my binoculars I saw not black-and-white birds spiraling into
the sky, but two Department of Environmental Protection officers lumber-
ing toward us.

Since the water from the Ashokan Reservoir glides south for over one
hundred miles to supply New York City, the reservoir and the land around
it are protected. We had permits from the DEP, which allowed us access

to the land in order to fish. We both carried a fishing license and a child's fold-up fishing pole. We were sort of following the rules. Except that we were not fishing. We were birding.

I turned my back to the officers. Their approach felt tortuous, like watching a spider inch toward its prey. My stomach twisted at the thought of being caught because if our trespass leaked out to the larger birding community, we would be blamed for tighter enforcement. I was only seven months into my birding life, but I understood that the local community was tight. And opinionated. We might ruin birding at the rez for others.

I glanced over my shoulder; the officers were fifty yards away. Looking away had not made them vanish. As I braced myself for the confrontation, I joined Peter in contemplating the far shore, where a dense hardwood forest bordered the reservoir. There the trees held onto their final leaves, the red oaks rippling copper-colored and brittle. A few yellows and reds—beech and sugar maple—shone, ready to drop. It was the end of October, the air sharp. The sky looked like winter, blue-gray and clear. The Catskill Mountains encircled us to the west and north, the steep rise of High Top gray-brown and imposing.

The weekend after 9/11, almost a decade before, I had hiked to the summit of High Top. There, I found several dozen hikers from New York City with bottles of wine and cans of beer, Zabar's bags, and good cheeses littered about. It was a confused gathering, suffused with a stunned exhaustion; what had happened to their city, to our country?

We sat in a loose bunch, gazing down on the reservoir, stilled in the aftermath of the attacks in the city. Every road in and around the reservoir had been closed, and would be for weeks. That hypervigilance had lessened over the past nine years. Now we were permitted to drive around the reservoir, and walk or bike across the breakwater. These freedoms included fishing, but did not allow for exploring the flats for birds. I understood why the DEP might not want people near the vulnerable water. Millions of lives depend on it being clean.

I have walked past my share of No Trespassing signs, often seeing them as more of a suggestion than a law. But this was well-patrolled and posted state land, and the moment we had walked past the long list of regulations I felt uneasy. It was only the lure of a special bird that had kept me going, made me feel like the risk was justified. Still, as we moved forward onto the flats, I felt spooked.

Maybe my unsettled feeling came from the land itself. In the early part of the twentieth century, ten thousand acres of land were cleared of two thousand people, ten churches, and five railroad stations. They even moved the dead, digging up graves and relocating them above water level, as they worked to provide water for Manhattan. It makes for a wonderful emptiness, though it's not a landscape that belongs here. It was my student Ali who first told me this history; the story she knew came via her grandparents or great-grandparents who had lost their homes. Years later, the family still resented their land had been taken. I didn't blame them.

If people are resentful of this man-made landscape, the Snow Bunting is not. The Snow Bunting breeds in the Arctic then travels south for the winter looking for similar flat, wide-open landscapes. Small and sturdy, as the bunting name implies, these pudgy sparrow-sized birds are striking in breeding plumage with a nearly all-white body, black eye, and stout bill. In this region of the Hudson Valley, though, I would never see the pure white of the Snow Bunting, as in winter the bunting's wings turn the color of the wind-whipped grasses.

I stood four feet behind Peter. I took in his wide shoulders and narrow hips; even in two layers of pants his legs looked skinny. A purple balaclava wrapped his thinning red hair. I wondered if he knew that purple wasn't his color. But I knew this wasn't a fashion misstep; he wore purple because it had been his son's favorite color. Peter continued to scan for the Snow Buntings while the officers approached. So I did the same, as if what we were doing was perfectly fine, utterly legal.

"They could be right there in the grasses, and you'd never see them," Peter explained. "The way to find them is to flush them."

Neither of us moved toward where the birds might be.

"Do we want to do that?"

"Don't worry. They always circle and land near where they started." Peter made this sound like a good thing, for the bunting and for us. Maybe it was.

Do we all flee what threatens us, then return to the same spot? I was thinking, of course, of love, how I have made the same mistake again and again, landing, fleeing, then circling back. The return to the same spot seemed a little stupid. Four months into my relationship with Peter, I wasn't sure, but maybe I had landed someplace new.

The DEP officers arrived, nodded a hello, asked for ID, fishing permits. No small talk.

Peter reached into his pocket for his permit; I did the same.

They held our fishing license while considering us.

What did they see? Two middle-aged white people, dressed for the cold with expensive binoculars strapped to their chests. Peter held an enormous camera with twelve pounds of lens, and I had a scope. Could they tell I reveled in the cold and Peter did not, that I had a talent for laughter that they would never hear? Would they care if they knew that Peter cried at romantic movies and that I couldn't sing in tune? Surely not. Only one thing mattered.

"What are you really doing out here?"

"Birding." Peter's voice was plain and honest.

The other officer nodded. "Seen the eagle?"

I looked at Peter and tried not to laugh. *The eagle.* We would be lousy birders if we missed it, but it's surely the bird people everywhere ask about if you have binoculars. It's an overloved and overadmired bird, for size and for what it stands for. And yet the poor eagle has such a pathetic call that in films it is often forced to plagiarize. More than once I've heard the call of a Red-tailed Hawk when the eagle soars across a movie screen. No one wants to think of our national bird as just this side of a crybaby.

"Not yet," I said too cheerfully.

"It nests just over there." He pointed into the distance.

We both tried to look excited before asking some questions about where we were allowed to walk. The officers admitted that the rules could be clearer but no, we were definitely out of bounds there on the flats searching for birds.

"What if we were fishing?" Peter pushed.

"You aren't." He tilted his head toward the pathetic poles that stuck out of our knapsacks.

One officer stepped away to call in a rowboat that floated in the distance. I hadn't noticed it before. Four men huddled together, the boat riding low in the water, not a fishing pole in sight. The boat wobbled slowly toward shore. I counted on the officers focusing on their new prey.

"Well, sorry," I said, moving away. I gestured for Peter to follow me. "Have a good day."

"Thank you," we said in unison, retreating to the edge of the reservoir, to the woods that lined the shore.

"How do you do that?" Peter asked. Peter had the sense I was an eel moving through the world, floating on warm currents and slipping past dams. Maybe I was part eel, letting life carry me along, always curious what the coast might offer, but I also know that eels are vulnerable when young, and when full grown are misunderstood: are they even fish some wonder. And, we all have dams we can never swim around.

I gave a half laugh.

Like dogs emerging from water, we shook off our encounter, grateful we had left without even a formal warning. What lingered was the disappointment that I had not met the Snow Bunting.

"We should go to East Kingston," Peter said when, the next Sunday, I announced I needed to see a Snow Bunting. East Kingston is a hamlet of a few hundred people, located on the northern edge of a city that was once the capital of New York State. Both the hamlet and the neighboring city show signs of economic struggle, houses that need a new roof or a coat of paint, or the car in the driveway that clearly will never run again. And both places also indicate a more prosperous past, the large nineteenth-century houses in the downtown section of Kingston or the beautiful stone and brick St. Colman's Church in East Kingston.

Where we would search for the buntings is the site of a former cement factory. "Stuff oozes out of the ground," Peter said. "The place is toxic. But if you want Snow Buntings," he said, his voice trailing off. I filled in the blank: Did I have it in me to suffer for the buntings?

Birding, I had learned, takes me to ugly places. Originally, I wanted to know birds so that when I hiked down a trail or kayaked the Hudson River, I would know what I was seeing and hearing. Soon birds became more important than my kayak or hiking boots and the beautiful places they led me. In the past few months, I had spent an odd amount of time at dumps, in the parking lots of malls or fast-food restaurants, and skirting water treatment plants. This was how I knew the birds had seduced me: I said *yes* to East Kingston.

It was a clear, blue-sky, cold November morning. The ravaged land where the cement factory once churned I found mesmerizing, drawn as I am to desolate, empty landscapes. Ruins of industry along the Hudson River tell stories of another world, one that used the river to ship bricks, cement, and ice.

We walked, Peter with his camera in hand, me with the scope slung over my shoulder, resting on the padding of my black down jacket. At the top of a rocky ridge we had a wide view onto a plateau framing the Hudson River. In the distance, on the eastern shoreline, trees stood near-naked in the early winter air. The Astor family once owned that land, a river separating them from the working, industrial, western side of the river.

Just past a cement tower, smeared with the sort of bold graffiti that appears on forgotten buildings, Peter stopped. "This is where I found my life Snow Bunting," he said. He cocked his head as if remembering that moment might call the birds to us.

Cedar Waxwings glistened in the bare trees eating the last of fall berries from a crab apple tree, and Eastern Bluebirds shuffled near the sumacs, still holding onto their deep red spear-shaped flowers. I wondered at the chance that a bird could make it from the Arctic to the Hudson Valley and intersect with my path. On this huge, great earth a bird and I were in the same place at the same time. The improbability of this brought on a giddiness as when contemplating distant stars and other galaxies.

For the next half hour we combed the spare land, walking a few steps, then stopping to scan, hoping for some movement, some sense of hidden life.

"These birds are holding out," I mused.

"That's because you're looking for a hard-to-find bird."

I liked the idea that this was a special bird, without understanding that specialness often has a dark side. In the past forty years, Snow Bunting populations have dropped by 50 percent. Researchers don't know why, and their work is complicated by the fact that the bird breeds in such remote places.

"Let's go down to that plateau by the river," I suggested.

Peter nodded. We walked side by side, our eyes still trained on the ground, scanning, hoping. Still, we didn't see them until the birds erupted from the earth, a fluttered explosion. Startled, we both stopped.

"Buntings!"

It took a moment for me to get binoculars to my eyes. The buntings circled around the empty lot, their flight buoyant, as much up and down as forward. Their jumbled movement looked a lot like large, tumbling snowflakes.

"Come on. Land," Peter coaxed.

I followed their movement, spinning on my heels to keep up with them, until my visual chase took my gaze directly into the sun.

"Damn." The stab of pain was sharp, as if I had been punched in the eye. I knew to stay away from the sun, but enchanted by the moment, I had given over to following the birds.

Through tearing eyes I could just make out the buntings as they landed about fifty feet from where we stood, not more than fifteen feet from where they originally foraged. In an instant, they vanished. This, in sparse grass no more than a foot tall. We moved slowly toward where the flock of twenty birds had settled. Nothing. They had vaporized.

"Where'd you see them land?" Peter asked.

I pointed to one spot, he pointed to another. The birds were not at either location.

"If you saw your life bunting here, and now I've seen my life bunting here," I started. Peter waited, perhaps expecting me to make some joke about how this was a sign we were meant to be together. I did feel that way, but a natural caution silenced me. After all, I was the rebound girl; Peter's second marriage had ended only a few months earlier. And I was nursing a heartache after years of loving a married woman. Summarized this way, our coming together sounds messy. But it wasn't. We stood side by side and looked at the birds.

"That means that the buntings come here, year after year. Yet before this was a cement factory weren't there trees here? This was not bunting habitat. Where would they go? And when did they start coming here?"

Several buntings rose from the grasses, fluttering to the sky like white angels. They circled several times, arcing loop-dee-loops, before swirling away over the gray water of the Hudson River as it flowed toward the sea.

"Good questions," Peter said as he took my gloved hand.

We walked past a dilapidated brick building back to the car and onto a second breakfast. I had only had a short time with the buntings, these messengers from the north, these sparks of winter. I wanted to spend more time with them, know more about them, and it unnerved me to think that this might be it, my only time with these Arctic birds.

LEARNING THE BIRDS

Woodstock and the Hudson River, New York

The path that led to standing in the cold next to Peter looking for Snow Buntings had started six months earlier, in the midst of spring migration. In mid-May, I arrived at Peter's house outside of Woodstock, New York, for the first time. I approached his house with great anticipation, as I thought of it as a hotspot for birds. He posted his sightings—Woodcock! Grasshopper Sparrows!—to the local birding list, which I read as if they were an almanac of the birds of Ulster County. His most recent post celebrated a Red-headed Woodpecker, a bird not common in our region, and an offer for people to come over to see it. I took him up on it.

A steep, pocked, gravel driveway led to his house, a modest log cabin set on a hillside thick with tall, white pines. He had made no attempt to landscape the property, maybe out of laziness, or maybe because deer and bear crash through, demolishing trees and decorative shrubs. The place was ordinary, just another house in the woods. It didn't exactly scream hotspot. What made it a hotspot was Peter. With a daily intensity, he combed the trees and walked down the driveway in search of birds.

That Peter found so many birds in this unassuming setting didn't surprise me. I had been on a few local bird walks that he had led, so I was familiar with his keen hearing and eyesight. I also knew that he conducted these walks with a daunting order: be quiet; don't be late. Some joked about being left behind in the parking lot; "the schedule said 6:30, we left at 6:30," he countered with a sliver of a smile. I took that smile to mean that he knew he was being a jerk, but also that he was right. As someone who is too-often late, I found myself arriving fifteen minutes early for these walks.

Once in the woods or in a field, I tried to stay near him, listening to how he spoke about the birds. In those words I heard intelligence and no sentimentality. His birding was energetic, and often full of humor. This countered the image I had long held of birders as being over self-serious, even pious. And yet he was clearly full of affection for the birds, often speaking of them as if of a friend. From the start I had a crush on him, obvious enough that another birder whispered to me, "he's married."

Married or not, he was welcoming me to his house to bird. As I stepped out of my car (two minutes late), American Goldfinches, like winged lemon drops, zinged through the air to and from a feeder suspended high in a tree. Next to the house stood a woodpile that impressed and concerned me: all of the logs, cut to the same length, rested in a too-neat stack. I thought of my own wood collection, short and long, some split, some not, which always looked on the verge of toppling over.

"Most people when they visit, I take them into the woods." He pointed behind his house. "Because it's you, you get to go to the dump."

The dump. I smiled like a teenager on a date.

The dump is the covered former Woodstock landfill, a few football fields worth of grassy land with white PVC pipes rising three feet out of the ground, letting methane escape from the rotting debris below. Because it is open land and empty of human disturbance, it's a great place for birds.

We walked down Peter's steep driveway, the crunch of gravel underfoot taking the place of conversation. At this late hour of the day, only an occasional bird song punctuated the air.

"Wood Thrush," Peter said without stopping. He motioned right, "Scarlet Tanager. Black-throated Green. Rumps."

Rumps. Rumps! I tried not to smile too wildly. The names of many birds had charmed, even seduced, me. I most relished those names that described the bird—Wandering Tattler (did it really wander?), or Mangrove Cuckoo (oh, to be in a mangrove). Surely a Whiskered Auklet was as cute as its name implied, and I wanted to hear a Fulvous Whistling-Duck whistle. Pairing birds and flowers—Magnolia Warbler—made the bird brighter, and any bird with *boreal* (Boreal Chickadee, Boreal Owl) in the name made me ache for the north. And I fretted over those birds named *lesser* or *least* (Lesser Nighthawk? Hardly), or those that were deemed *common* (Common Swift? Nope).

Once I entered the bird world, another layer of bird names unfolded for me, as I coveted the casual way that birders tossed back and forth birds they had seen in a sort of shorthand, one that hinted at a loving intimacy: "Did you see the Hoodie and the Buffie?" Once I heard these affectionate diminutives, I wondered who wouldn't want to get up before dawn to see a Maggie, to greet a Snowy, Swainy, or Swampy? Now here was *Rump*, short for Yellow-rumped Warbler, a bird someone on our last walk had called a Butter-butt.

We crossed a road and entered the gravel roadway that led to the dump.

"Let's stop at the marsh," Peter said, pointing to a faint path that led off behind a large red maintenance building. "The woodpecker has been coming in at dusk; maybe he'll be early."

We brushed past dense multiflora rose that snagged at the edge of my pants, and after a few yards we stopped at the edge of a derelict pond, dotted with tall snags. In patches the still water glazed with a milky sheen. A belch of methane, the soft rot of life, drifted over us. I admired a Great Blue Heron that posed, awkwardly statuesque, its long feathers a wispy, messy shawl. A pair of colorful Wood Ducks disappeared into the reeds. It took a few months of birding for me to see such places as beautiful because of the possible birds that might be drawn to a swamp. At that moment, new to birding, I was cheerfully disgusted at how uninvitingly fetid the place was. Perhaps there was a reason others hadn't jumped at Peter's offer to see the Red-headed Woodpecker.

Peter extended the legs of his scope and set it up in the damp ground. We stood a foot or two apart, binoculars at the ready. I shifted my weight from one foot to the other as an electric quiet enveloped us. Minutes passed. I wondered if it was all right to talk, but didn't risk it.

"You need new bins," he said, breaking the silence.

I nodded, self-conscious. I had bought these binoculars at the Audubon shop when I had been living in Tucson, Arizona. When I tried them in the store, I focused on a sign hanging outside a window across the street: "Stop looking in my window." I had marveled over how crisp the letters appeared and quickly shelled out $250 for my new binoculars. At the time I was a graduate student so they felt extravagant, a luxury, yet compared to Peter's optics they might as well have been a plastic gift found in a box of Cap'n Crunch.

"How long have you been birding?"

"Five years," he said. At the time, he was in his early forties.

That gave me small hope that I might not always feel so lost.

"Think of it this way," Peter said. "There are around two hundred birds in this area. That's not a lot to learn."

I translated this into a familiar domain. Two hundred French verbs. That was doable. Yet when to reach for the subjunctive, or how to distinguish between all of those sparrows was tricky. And the truth is, when to reach for the subjunctive was intuitive for me, as I had learned French the easy way, by having my mother speak to me as a child.

Unlike kids today who are seen and heard and take up the entire living room, my sister Becky and I grew up with adults living by the proverb that "children should be seen and not heard." So we were silent as we sat in big chairs while the adults drank their martinis and talked. Oriental rugs, a heavy portrait of a long-dead distant relative on the wall, inherited lumpy couches that needed to be reupholstered, shelves of books lining most walls: ours was a classic academic, literary house, rich in the overstuffed and the under cared for. Not shabby chic, just shabby, comfortable. If something broke my mother hunched her shoulders in a so-what gesture. Things didn't matter; ideas mattered, books mattered. In this setting, listening intently and asking questions when I could, I learned about books and ideas, and my mother made sure that I learned to speak to her in her language, French.

I wanted to learn birds in this same way, by osmosis. I decided I would range through the woods and streams, and learn the birds by keeping ears and eyes open, by paying attention. And yet I also knew that going out on my own was not productive, that I spent too much time baffled by even the most common birds, and was often stumped by how or where to look. I needed to become familiar with the common local birds before I could venture out alone. Finding the local bird club, named no less for a once famous nature writer—John Burroughs—felt like a breakthrough. It was filled with the oddballs I had anticipated, but I also saw that these folks knew their birds and some were generous with that knowledge, didn't mind waiting while I located the Yellow Warbler, which sat out in the open and sang with glee. That I seemed to have little natural aptitude for this new passion made me want it all the more.

Like many people, until I became a birder I was not familiar with Alexander Wilson, who arrived on North American soil in 1794. He is the author of a nine-volume work, *American Ornithology*, and considered by many to be the father of American ornithology. In his *American Ornithology* Wilson writes of the Red-headed Woodpecker: "There is perhaps no bird in North America more universally known than this." Boldly colored, red, white, and black, "glossed with steel blue," it is a bird hard to miss. Every orchard and cornfield held a Red-headed Woodpecker helping itself to a meal. Every child, Wilson claimed, was acquainted with the bird.

It amazed me that I had never seen this bird every child, at least in the late 1700s, once knew. Red-headed Woodpecker populations have gone up and down over the decades, at one point nearing extinction. Since the midsixties, according to the North American Breeding Bird Survey, populations have decreased by 70 percent.

From the start of my bird journey, learning these sorts of dire statistics made me feel as if I was perched on a precipice. One step; one less bird. I didn't feel despair, rather dizzy with disbelief and overwhelmed with an urgency to see the birds. This urge was part selfish desire, but not entirely. Hooked into my sentiment was a sense that my witness of the birds could—would—keep them tethered to this earth.

This bird that Peter and I hoped to see is considered by some to be Wilson's spark bird. When Wilson arrived in Philadelphia, from Paisley, Scotland, the birds in this new country caught the young poet's eye. In Philadelphia, Wilson had the good fortune to meet William Bartram, who was to become a famous botanist and ornithologist, as well as William's father, John Bartram, who was already a star in the botanical world. John Bartram's work filled Wilson "with a kind of creative enthusiasm." At every turn, Wilson's joy and fascination in the natural world is what pushed him forward. Eventually his curiosity for the birds led him to a vision of his ornithological guide, which he imagined in ten volumes.

What a world that must have been, every bird a new discovery (of the 320 birds he describes in his *American Ornithology*, fifty-six were then new to science). The challenge of this is hard for me to imagine: finding the bird, shooting it (this I can't imagine), then drawing it (this, too, is beyond me). To create a bird guide in the early 1800s required enormous energy, more than just up-at-dawn, as Wilson moved around his new country, day after day, on foot or by horse.

Wilson travelled restlessly, referring to himself as a "bird of passage," looking for new bird species and also for subscribers to his work. He needed to fund his passion. At one point Wilson journeyed 720 miles down the Ohio River in his *batteau*, a flat-bottomed boat, which he had named the *Ornithologist*. On board, he had a tin cup "to bale her, and to take my beverage from the Ohio with." Imagine that, to drink directly from the Ohio. I envied this cleaner world and the wealth of birds that came with it. At one point in Kentucky he calculated that "two thousand million Passenger Pigeons" flew over one afternoon. Two billion birds. What an ocean of birds! And now, all of those billions of Passenger Pigeons are extinct. In a field one day Wilson marveled over a thousand Clapper Rails. To see one or two of these secretive marsh birds would seem to me a treat. So much of Wilson's world is gone, like his pet Carolina Parakeet, which he carried in his pocket from Big Bone Lick in Kentucky to New Orleans. His beloved Carolina Parakeet is an extinct species as well.

Wilson was at times overwhelmed by his own obsession. He complained of having no time for friends, or leisure. "Even poetry, whose heavenly enthusiasm I used to glory in, can hardly ever find me at home, so much has this bewitching amusement engrossed all my senses." *Bewitching*. How quaintly sexy, like Ella Fitzgerald singing *bewitched, bothered, and bewildered am I.*

I liked what I read about Wilson—poet, teacher, romantic, political agitator, and bird lover. Was Peter any of these things? I only knew that he loved birds. Still, I was a little bewitched by him as we stood a little too close and waited for the Red-headed Woodpecker. And waited. I realized from the start that the birding characteristic I would need to cultivate was patience.

I associated patience with slowing down, with sickness—patients require patience—and, I had to admit, with dullness. I like moving: energy begets energy. Moving helped me to outrace age, boredom. But there was something different about this birding patience, something stoic in our perseverance—how long could we stand there—and I sensed energy in our quiet waiting there by the pond, a tension that came from paying attention.

In the now dimming light, I wondered what we might see, or if we would see the bird at all. I tried to stand straight as my lower back, with growing discomfort, put in the case for quitting soon. How hard it is to stand still. Every time my thoughts started to drift, I forced my gaze back to the pond, my eyes soft focusing on a snag like a newcomer to meditation. I was just settling into a still point when out of the woods shot the bird, all wings and a raucous bluster.

Movement. Peter snapped to the scope, focused, then, without lingering at the eyepiece, stepped back and gestured me toward the tripod. It took me a moment to adjust to peering through one eye, finding the focus wheel and bringing the bird in close.

"Oh. Wow."

A magnificent vermilion head met with a black body and a white chest. The bird was stocky, like a boxer, the colors sharp. It would be an easy bird to draw, the lines between red and black and white crisp. The colors themselves would be the challenge, the black tilting toward silky blue and the red so deep it suggested royalty. This bird was a prince, regal yet filled with a youthful energy.

"Gorgeous," I whispered, more to the bird than to Peter. It is possible, when looking through a scope, to feel as if you have fallen into the bird itself. My eyes brushed each feather, took in the bird's dark eye and its strong X-marks-the-spot feet that allowed it to cling upright, its belly pressed against the trunk of the snag. It became a part of the tree, a stumpy colorful branch.

"Do you think it will stay?"

"I hope so," he said. Something about the way he said this made me feel Peter's longing for the bird but also for something more, some bigger wish. "But I doubt it."

"What would convince it to stick around?" Head bent over the eyepiece of the scope, I could feel the sun setting. I wanted this bird to stay, not just for the day but the season, to find a nest hole and produce baby Red-headed Woodpeckers.

"A mate." Peter stated the obvious. He paused for a moment. "The chances of that are slim."

"Really slim," I said too quickly. What did I know? In all of the hundreds of thousands of acres of woods this bird had to find a girl (or boy—male and female Red-headed Woodpeckers look alike) with a red head and a black cape with white trim. *Impossible*, I thought. Surely I was imposing my own life onto this bird. But there I was, near fifty and single. The older I got, the more astonishing couples seemed: the woods are dense, the swamps unending. But maybe it was simpler than I thought; maybe I'd just been looking and waiting in the wrong dump.

Later, when I told my friend Teri about seeing a Red-headed Woodpecker, she laughed, "That's a great pickup line: *Want to come over and see my Red-headed Woodpecker?*"

I laughed a little too hard at her joke, bewildered as I was by the combination of Red-headed Woodpecker and married man Peter.

A week after the Red-headed Woodpecker, Peter sent an email. "I meant to ask you, if you are ever looking for company out on the river, I always love an evening paddle on the glassy Hudson as it's getting dark. I don't get out there enough."

The Hudson is rarely glassy; it's a big river, tide and wind torn. If Peter were a student in my nature writing class, I would have circled *glassy* and written *word choice?* in the margin. However, Peter was given extra credit just for writing, all clichés forgiven.

The night we went out at the end of May, wind blustered the river into cold whitecaps, as the wind and tide battled for dominance. Peter kept twenty strokes ahead of me as we pushed our sleek sea kayaks south on the river. The unpredictable movement of the water forced me to focus on each stroke, not on getting to know Peter.

A half mile south of our launch in the village of Tivoli, Peter waited near the railroad underpass, his paddles swirling like he was treading water. "What's in there?" he asked, gesturing under the bridge toward the North Tivoli Bay.

"Let's go in," I nodded. It was fun to be showing Peter my world, the Bay where I spent mornings and evenings checking on the beaver, searching for golden club in spring and cardinal flowers in fall. I knew the Hudson River and the Bays well—the history and the texture of the land. But there was a big gap in my knowledge: the birds that brought this world alive, that sang while I poked about, greeting snapping turtles. Beyond the exuberant Red-winged Blackbirds there were few birds I could identify.

We dipped into the tunnel, the rusted metal supports just inches from our heads, our paddle strokes constricted by the low bridge. Once inside the Bay, the water opens out like an inland lake, rimmed by cattails, phragmites, and wild rice that looks exotic as it waves in the wind. Somewhat sheltered from the wind, we were able to move side by side, our paddles nearly knocking in slow synchronicity. We trolled down the wide watery pathways that meander through the Bay, the song of Red-winged Blackbirds punctuating the air with their *conkla-dee!* Peter pointed toward the staccato song of the Swamp Sparrows, then the electronic cascade of the Marsh Wrens. *So that's what that is!* I thought, the songs so familiar.

Our conversation meandered like the waterways of the Bay. His parents had died, as had mine. He was the youngest of six siblings, while I had only Becky, living across an ocean. He offered the not entirely unexpected news that he was mid-divorce. And he told me, matter-of-fact, that he had had a son who had died.

"Oh." I felt a jolt, as if I might cry for this child I had never known.

"He was three." And then he gave me the name of the too technical and nearly unpronounceable form of cancer that had taken his life.

I had lost parents and a cousin, an aunt, some friends. Cancer, heart attack, drowning, suicide. I knew how clumsy people were in how they responded: *I'm sorry, my heart goes out to you.* There were no words for a man who had lost his son. I felt grateful for the steady beat of our paddles, something to do when I had nothing to say, except to ask him his name.

"Forrest."

How perfect. I listened as Peter spoke of the too-brief time that he spent with his son, not as if the boy were still alive, but rather that he lived inside of Peter.

It was five years ago, after Forrest died, that Peter bought his first pair of good binoculars. With his new Zeiss binoculars in hand, he stood on his porch peering into the trees to see what he could see—he wasn't even really looking for birds—when he spied something moving. As he focused his bins, a Blackburnian Warbler filled his eyes.

The Blackburnian! I had never seen this special warbler, but I knew its reputation as one of the most striking birds with its fiery orange chest. Peter's response when he saw that vibrant orange was to wonder how he had not seen this bird before. He had grown up outside, scooping frogs and turtles out of creeks, pulling snakes out from under logs. He hunted squirrels and white-tailed deer near his home in southeastern Pennsylvania. He knew the woods. Yet since the Blackburnian was news, what else might be out there? At that moment, he set out to know.

As Peter told me this story, we followed the water path through the Bay as it curved to the south, edged on one side by a stand of phragmites and the other by land, oak trees that draped over the water. Our kayaks parted the waters in a simple V, the long view before us giving me the sense we could paddle to the edge of the world. We moved like this in silence until Peter started telling me about the time he joined the push to find the Ivory-billed Woodpecker in Arkansas. My mind hiccupped: *he went searching for a bird thought by many to be extinct!* He described paddling through the swamps of the White River and Cache River National Wildlife Refuges locating himself with a GPS in the tangle of water trails, eager to find this mythic bird. The isolation, the raw natural beauty—I jangled with envy. Then he described the water moccasins that course the waters of those swamps or dangle out of low hanging branches, and I wondered if I would have the courage to venture there. I'm a fan of snakes, but not at close range.

When in 2004 Tim Gallagher, writer and long-time editor of the Cornell Lab of Ornithology's magazine *Living Bird*, reported he had sighted an Ivory-billed, hope washed through all of us who pay attention to the natural world and the slow drips of extinction. I had read about the months then years of searching, and how the Cornell Lab of Ornithology rallied the best birders in the country to comb the southern swamps armed with cameras and recording devices. No one was able to get a confirming photo or recording.

"How did you get to go?"

"I lied." At that time, in 2007, which was toward the end of the organized search, Peter had been two and a half years into his birding life.

At night, Peter had slept in his work van outside of the "almost abandoned town of St. Charles." His spare, solitary life extended for ten perfect days as he held onto the hope he would find the bird, would chance upon this big woodpecker that alights in the dreams of many.

My father was a storyteller. His tales and jokes animated our evening meals. One tale he told, over and over, involved two little boys, one an optimist, the other a pessimist.

The parents of these two little boys fretted that the optimist was just too cheerful and the pessimist simply too glum. So they brought in an eminent psychologist, who spoke with the appropriate German accent, to diagnose and then prescribe what to do. Stick the pessimist boy into a room with every toy you can buy, the psychologist recommended. Leave him there for twenty-four hours. For the optimist, he determined, confine him to a room filled with horse poop. (The word *poop* made Becky and me giggle madly like the little girls that we were.)

The next day the parents walked down the hallway and could hear sobbing coming out of the room of the pessimist.

"What is wrong?" the parents asked in frustration, when they opened the door. "You have every toy you could imagine here."

"Yes, but if I play with them, they will all break," the pessimist responded.

The parents shook their heads, discouraged, and continued on to the room occupied by their little optimist child; the squeals of joy reverberated through the house. They opened the door to find him tossing horse manure into the air. (*Manure!* More gales of laughter from Becky and me.)

"What is the matter with you?" they asked. "You are in this stinky room filled with horse poop. What is there to be so excited about?"

"With all of this horse poop, there must be a pony here somewhere," he responded.

The first time I heard this story I surely didn't know what the words *optimist* and *pessimist* meant, and yet I laughed—as I laughed every time my father told this story. Over the years, I came to understand the words and more: people are who they are.

In my world, optimists are suspect, too cheerful so perhaps naïve (who can possibly remain happy once they have read Locke? Rousseau? Nietzsche?), while pessimists are considered the hard thinkers, the ones who see the awful world as it is, and awful people for who they are.

I have tried to discourage my inner optimist, if only to fit in with the world that I live in. But I was always the one who was sure there was enough time and kindness, enough goodness and fairness, even enough of that rare element, love. I have never stopped looking for the pony amidst the poop.

In the bird world I learned there's an easy test for optimism: ask if there are Ivory-billed Woodpeckers cackling away in southern swamps. Pessimists will say no; optimists will say yes and even go and look for them. Here was a world where my life-long optimism could be an aid, even a virtue, encouraging me to go out and find the impossible. The bird world would allow me to accept who I was, a person who lives her life full of hope.

BICKY

Slide Mountain, Catskills, New York

In the hushed light of the Catskill Mountains, the dawn chorus swirled and mixed into a lispy symphony. From time to time, out of the underlying burble, a note rose higher, louder than the others. What I heard was a celebration of life.

Mark, one of the leaders from the John Burroughs Natural History Society (JBNHS), stood in the gravel parking lot off of Oliverea Road and pointed into the dark of the cool June woods: Ovenbird, Redstart, Black-throated Blue, as if separating out the notes of a trumpet or an oboe from a full orchestra.

Mark was young, gregarious, and generous. He made jokes—most of them making fun of himself—and had a trickster's way of being able to talk and spot birds at the same time. He carried a guidebook wedged between his pants and his butt and pulled it out from time to time to be sure everyone understood what they were hearing and seeing. The bumper sticker on his pickup truck read: Take a Kid Birding.

"Let's go," Mark said, gathering our attention. Five of us tightened hiking boots, tossed on knapsacks, wondered out loud if we'd need headlamps to trek up Slide Mountain, famous as the highest peak in the Catskills. I didn't know any of my fellow hikers and was disappointed to see Peter had not shown up.

"He's sleeping in," Mark said. Several joined me in laughing: Peter did not sleep in.

I knew well the trail up Slide Mountain. I had hiked to the summit of Slide in all seasons, and often in winter: layers on, hat in place, water and snacks at the ready. On those hikes, the summit was the goal—I admired the view, then took some photos that I always soon deleted because nothing can capture the magnificent sense of space at the summit of a mountain. On this hike, the summit was not our goal: we were hoping to hear the Bicknell's Thrush sing its morning song. Suddenly the mountain seemed so much more interesting to hike up, the summit holding a bird that would surpass the views and be the main event.

That a bird was partial to the summit of Slide made it pretty special but what an odd, even challenging, place to set up shop. For the naturalist John Burroughs, this choice in nesting location not only made the bird interesting but enchanted the mountain itself: "Slide Mountain enjoys a distinction which no other mountain in the State, so far as is known, does,—it has a thrush peculiar to itself," he wrote in his 1910 essay, "Heart of the Southern Catskills."

What Burroughs called the Slide Mountain Thrush is now known as Bicknell's Thrush, named for Eugene P. Bicknell, a nineteenth-century New York man, both botanist and ornithologist, who made his living as a banker. He had an excellent ear, and in 1880, on the summit of Slide, Bicknell heard an unfamiliar song. For a time, the bird was considered a sub-species of the Gray-cheeked Thrush, but now stands alone.

The Bicknell's is fussy about altitude, but researchers have affirmed it is not so fussy as to live only on Slide. It breeds on several peaks from the Catskills to northern Maine. What it does need is to be above three thousand feet, and it prefers balsam fir habitat in areas disturbed by wind or winter ice. Provide these particular conditions and the Bicknell's is happy.

Before we reached the summit, before we might hear the Bicknell's, there were so many other birds. We stepped gingerly on rounded rocks across a stream and started up the uneven carriage road that ran for a little over half a mile southwest, then more steeply uphill and east for two miles to the 4,180-foot summit. I walked behind Mark, hoping to learn from his keen ears as he pointed to bird songs, "Wood Thrush," "Winter Wren." He paused briefly and pointed up, toward a high-pitched song that soared toward the stratosphere, a song that I could easily have missed. "Blackburnian."

I wanted to track down the bird as I had not yet met the Blackburnian, the bird that had launched Peter's bird life.

"We have to keep moving," Mark said. We needed to be on the summit early, to catch the Bicknell's singing its dawn song. "You'll see a Blackburnian," he said, as if the woods might be full of them.

Blackburnian, I mused, the clunky name leading me to wonder if there might be an intimate diminutive (Blacky? Burny?). I was not alone in thinking the bird deserved a better name. Burroughs wrote that "the orange-throated warbler would seem to be his right name, his characteristic cognomen; but no, he is doomed to wear the name of some discoverer, perhaps the first who rifled his nest or robbed him of his mate,— Blackburn; hence Blackburnian warbler." Burroughs is correct, the Blackburnian is named for the brother-sister team Anna and Ashton Blackburne. Ashton traveled in the States, sending home to England bird specimens for Anna's collection. One of these birds was a warbler, shot in some unspecified location in New York State. The new bird specimen was given to the naturalist Thomas Pennant, who described and classified it, and finally named it in honor of the Blackburnes.

There are 149 birds "doomed" to carry the name of a person. Some of these people are the original discoverers of the bird, like Townsend, Kirtland, Swainson, Cassin, and others. Some were supporters of the ornithological world, like Henslow, Rivoli, Le Conte, or Smith. Having these names introduced me to individuals and their histories and also to a larger history of exploration (more western birds hold eponymous names than eastern birds). But as I read more, part of me started to feel that many of these people did not deserve such a fine monument, and that for the birds the names might be a bit of a burden. A thrush should not have to live forever associated with a banker ornithologist.

And, of course, the naming is skewed toward white men. It took some digging, among both common and Latin names, to find eight birds

named for women. (There may be more, but this is where my searching landed me.) Eight. Among the seven-hundred-plus species in North
America. And not one is named for a woman who was an ornithologist,
but rather all are a sister, daughter, wife, or (financial) supporter of an
ornithologist.

John Burroughs was born in Roxbury, in the Northern Catskill Mountains, in 1837 and spent his life exploring the hills and valleys, the birds
and porcupines of the woods. Through the course of twenty-three books,
the Catskills are central both geographically and emotionally to Burroughs's work. During his time, he was a celebrity, sharing hiking trips
with John Muir and driving a Ford Model A given to him by his friend
Henry Ford. People eagerly read his work. But Burroughs's popularity has
not endured. Though few read him, I do, and perhaps my favorite essay
is "Heart of the Southern Catskills" where he describes the three days he
spent hiking and camping his way to the summit of Slide. This was his
third attempt to get to the summit of what he called a "shy mountain,"
one that was not giving up its secrets or its summit easily. He started his
ascent in Woodland Valley, where "only men may essay the ascent." I like
Burroughs's use of *essay*—from the French *essai*, to try—a verb somewhat
lost in the shuffle of time. The noun remains in its many forms: creative,
five paragraph, collaged, research, personal. The essay is what I write and
what I teach. Noun or verb, it's about trying, to get the right word, to get
to the top of an idea.

What I don't like is Burroughs's sense that only men can ascend Slide
Mountain from Woodland Valley, a hike I had made several times. On
my hike with the group from the Burroughs Society we were taking the
shorter, more direct route up the mountain, one that Burroughs claimed
was "comparatively easy, and where it is often made by women."

Burroughs was a man of his time—shooting birds to identify them,
"playing with" (or rather torturing) a porcupine on his ascent of Slide,
and on another excursion in the Catskills ("Birch Browsing") chasing off a
mother Ruffed Grouse so he can slip one of the babies into his coat sleeve.
(It then runs up into his armpit, which is sort of endearing but also not
OK.) His attitude about women is also of his time: women can make it up
Slide, but only the easy way.

I shoulder my way through these unpalatable comments in order to get to the reason I read Burroughs: how he writes about birds. Few descriptions of birds are as eloquent or original as his. Here is how he wrote of the Bicknell's Thrush: "The song is in a minor key, finer, more attenuated, and more under the breath than that of any other thrush. It seemed as if the bird was blowing in a delicate, slender, golden tube, so fine and yet so flute-like and resonant the song appeared. At times it was like a musical whisper of great sweetness and power." This was what we were listening for as we moved toward the summit.

Our route did not resemble Burroughs's meander up the slope past streams "soliloquizing in . . . musical tones . . . amid moss-covered rocks and boulders." What we were doing in a few hours in a morning took Burroughs three days of route finding. Our ascent required no imagination or risk as we followed the worn, rock-strewn carriage road that shoots upward at a steady incline through a hardwood forest of red maple, yellow birch, and American beech. As we marched upward, I learned from Mark that thrushes frequent the shaggy woods of Slide, their elaborate songs emerging from the deep, speaking of dead leaves and fallen limbs, of hidden hollows. Each thrush has its own altitude: the Wood Thrushes down low, the Swainson's and Hermit midmountain, while the Bicknell's claims the top. On the airy summit is where the Bicknell's breeds.

The entry in Birds of North America online offers that the Bicknell's is one of the "least known breeding birds in North America." Much of the information we do have rests on a study by a Vermont farm boy named George Wallace who wrote his PhD dissertation on the Bicknell's Thrush at the University of Michigan in 1939. Wallace conducted his study on Mount Mansfield, Vermont, beginning in 1935. In Wallace's memoir, *My World of Birds: Memoirs of an Ornithologist*, which is a fun read, he describes how he and his young bride spent the summer high in the mountains living in a cabin that served as his research station. There they ate porcupine stew (which they found tastier than squirrel stew) while during the day they searched for Bicknell's Thrush, finding many birds as well as twenty nests in trees near the summit. They rescued one young thrush, making it a pet that lived twenty-nine happy days until it accidently plunged into a pot of boiling water. They took in a second bird, bringing it home to Ann Arbor. There, "Bicky" provided "a constant flow of thrush like musical notes." The following year, they returned to Mount Mansfield and released Bicky.

How wonderfully not scientific all this is! It's hard to imagine a bird researcher today writing about making their subjects into pets. Besides the fact that this is unethical, they would no doubt be shunned within the ornithological community. And yet when I see photographs of researchers grinning wide and holding banded birds, it often looks like they want to pet and hug them, want to take them home.

Something about birds as pets eclipses me. It's not just that the mostly featherless, hairless legs and sometimes talon-bearing feet are far from cuddly; it's that I wouldn't want to domesticate an animal that wants to be wild. What I find most inspiring about birds is how free they are, winging off to the top of a tree or sailing over the horizon. To want to tame that, to hold it, to bring it indoors seems a bit like wanting to bring home a waterfall.

Having Bicky near no doubt made Wallace fond of his subject, determined to further knowledge about his beloved pet bird. And yet, it is a bird we still know little about, which is rare for a North American bird. At first I imagined that the reason it was understudied was because of the remote, high elevation and dense forest habitat where they breed. Researching a bird so hard to access must complicate long-term studies. But the biggest challenge is that Bicknell's is a bird that confounds researchers with its mating habits. Both males and females have multiple partners, multiple paternity is common, and more than one male often feeds young at a nest site. That's the sort of sentence that is hard to follow. Try following the actual birds as they conduct these complex lives. Put another way: only 59 percent of males actually sire the broods that they feed. It sounds chaotic. Yet all I can think is how creatively not monogamous all this is.

The Bicknell's is a model of alternative living (no swans these birds), proof positive that raising a brood collectively works, or that multiple partners—a winged version of the open relationship—doesn't require group therapy. In fact, it's a cheerful arrangement for all. Wallace observed that the female sings at the nest during incubation, hatching, and brooding. Particularly during the hatching process the female sometimes "perched on the rim of the nest and sang—in apparent ecstasy—while watching the eggs hatch." All of this happy nest and mate swapping makes the Bicknell's, in my mind, a pretty cool bird.

Yet no one seems to be praising the Bicknell's happy ways. Mating habits are most often noted when animals mate for life. "Awwww," people say, as if there is something romantic about geese sticking together. But the decision to be monogamous is not a decision as we know it. Being monogamous is done out of survival, because it's efficient or safe, helps the bird breed successfully in a dangerous world. Birds that are monogamous are so because it works for them. And the Bicknell's is nonmonogamous because it works for them. They are not (except in my imagination) the swingers in the woods.

Near the 3,500-foot elevation point, the trail turned sharply left, narrowed, leveled out, ascending gradually as it ran through a forest of balsam fir. This was where we started to listen even more keenly for the Bicknell's. We walked through what felt like a green tunnel, the trail darker as the trees haunted together. The rocky path gave way to a bed of needles, my feet sinking as if kneading the belly of the earth. With each muted step, the morning air misted my skin and the fir scent opened my senses. Bird song pitted the air: Dark-eyed Junco, White-throated Sparrow. We had been walking now for a little over two hours, and in that time the sun had risen but wasn't yet warming the air or fully lighting the dense woods. In the coming light, I caught a glimpse of yellow or a dash of black and white. Mark called out: Magnolia Warbler, Blackpoll. A *Maggie*. Yellow with black stripes at its chest, it looked both cheerful and determined, an intrepid bird living high in the mountains.

When the song of the Bicknell's emerged, it took me by surprise, circling like a delicate musical force out of the dark woods. We all stopped, breathing heavily from the steady climb and from excitement. We stared through the skinny tree trunks, into the vibrant green ferns and moss that carpeted the woods where a fragment of light dropped to the high-altitude forest floor. Nothing looked like a bird. The sound pierced the brisk air, a deep, chocolate-rich whirl.

Mark *pished*, which sounds like a loud, aggressive hush, imitating the alarm call of a Tufted Titmouse. The hope was that a curious Bicknell's would come in to see what the fuss was about. We stood single file on the mountain trail, all scanning the woods for this thrush of medium size, dusky brown, with a spotted breast and mournful eye. From time to time someone would lift his or her binoculars and we'd all wait, expectations high, for a "got it," or "there's the bird," or simply, "yes!" At any movement, I would rush to focus my binoculars, but each time it was a Purple Finch or a Black-capped Chickadee blazing by. I waltzed from being discouraged, then excited, then frustrated as others called out their sightings, fun birds all but not Bicky.

A Bicknell's never appeared. It was as if a phantom was vocalizing from deep in the woods—all sound, no body. Was the song enough to say I had met the Bicknell's? It isn't an idle question. Saying you have *seen* a bird is a claim, one that the American Birding Association has weighed in on. In 1992 the ABA changed their rules on what a birder can list, known as the countability rules; heard-only birds count. *Countability rules.* That rules had to be made told me all I needed to know about how seriously everyone takes their claims of seeing or hearing a bird, and the accuracy of their list of birds. I decided that for my first meeting with the Bicknell's I needed to see its lonesome eyes.

I did not feel disappointment as after an hour and a half of searching we accepted defeat, but rather a sure determination to return. The Bicknell's is a rare and threatened species, one that the 2007 Audubon Watchlist placed in the Highest Continental Concern (HCC) category. The Audubon Watchlist (which includes 176 species of birds) notes that there are forty thousand individual Bicknell's worldwide. The Bicknell's suffers from mercury poisoning on its breeding grounds, while on its wintering grounds in the Dominican Republic, Haiti, Cuba, Jamaica, or Puerto Rico it finds its greatest challenge: habitat loss and destruction.

The more I read about birds, the more I read the lines *habitat loss and destruction.* This refrain is shorthand for all the many ways we have plowed over, plowed up, and plowed down this earth; all of the ways we have poisoned our waters, our soil, the skies; and all of the ways we have made a mess of this great planet we call home. So the pursuit of the Bicknell's, like that of so many birds, is necessarily an exercise in patience and an act of hope.

It was hard to feel any of that loss, though, as we loitered on the summit for a while looking out on rolling waves of green that were nothing but inspiring. We posed as a group in front of a plaque commemorating

Burroughs's ascent of the mountain. There, a quote from his essay "The Heart of the Southern Catskills": *Here the works of man dwindle*. Taken out of context, it might seem that Burroughs is emphasizing our smallness when in fact he is celebrating the bigness of the world, how from the summit, "the original features of the huge globe come out." The Hudson River is "only a wrinkle in the earth's surface. You discover with a feeling of surprise that the great thing is the earth itself, which stretches away on every hand so far beyond your ken." What he sees from the summit is the round earth itself. And the earth he saw was not yet suffering from too much destruction.

Burroughs was a nature writer prone not to awe, but rather to intimacy. So that moment on the summit of Slide is less about feeling overwhelmed, and more about feeling closer to the earth, *because* he can see her contours. This ability to bring the big to an intimate scale is one of the reasons I am so drawn to Burroughs's work. I craved that intimacy with the natural world and saw how birds gave me that. And yet I also still needed the rush of the big, the cold, the vast. Slide gave me both.

On the descent from Slide, not more than a mile from the parking lot, my knees slightly aching and my shoulders pinched from my knapsack, long after my mind melted soft with bird sounds and I had given up trying to distinguish what still sang in the late morning hour—the bird appeared, the Blackburnian I had so wanted to see at the start of our walk. Perched in a tree just above me, the chest glowed even more vibrant orange than in my imagination. I didn't need anyone at my side to identify this bird for me.

The bird tilted his head as if looking at me, its dark eye framed by a black eyeline that cut into his orange throat. The bird's spindly legs clung to a narrow branch, the light filtering through the trees highlighting first the yellowish belly, then the white wing bars, then again that dazzling orange throat. *What a design*, I thought. I understood why Peter had fallen for this bird.

As the bird flitted about above me, I thought, *how true this is in life*. After we stop looking, there's the Blackburnian.

METHINKS

Mount Desert Island and the Golden Road, Maine

The Nelson's Sparrow is a solitary, secretive bird with a sliver of orange above the eye and a weak, airy song that has been described as a "hot iron being put in a bucket of water." A beginning birder should not be looking for a Nelson's. I needed strong, obvious colors and a bird willing to sit up while singing. But Peter wanted this bird, and I was happy to tag along as the quest for the Nelson's felt like an initiation into a secret society made up of those willing to sort through the little brown birds that most people, and many birders, ignore.

Our search circled around Mount Desert Island, Maine, where Peter and I had landed on our first bird trip together. On a morning walk a few weeks earlier, just after hearing a Ruffed Grouse drumming at dawn, the beat mellow and strong as if the earth itself were pounding out a love song, we held hands. Our new romance, bleary with before dawn starts and, for me, with end-of-semester busyness and, for Peter, with his steady work as an electrician, had left us looking for a quieter time together. We decided on a little vacation, which was easy to organize as I had the summer off from teaching, and Peter works for himself. So we strapped kayaks to the top of my car and stuffed tent and sleeping bags in the trunk and headed north and east to see new and different birds.

At night we studied maps in the light of our headlamps to locate the salt marshes where Nelson's live and breed. We spoke of the bird as if looking for a long-lost cousin, referring to it as Nelson. While most visitors to Mount Desert explore the dramatic, rocky coast, stare out across the ocean, we focused on streams and marshes in the interior, secluded places mosquitoes call home. Ankle deep in mud, I reveled in the late June dawn chorus, now able to pick out many of the warblers I had just come to know, the *witchita, wichita, wichita* of the Common Yellowthroat, the ascending *ziipppp* of the Parula, the high, thin *swee* of the Blackburnian, and the so sweet Yellow Warbler. How fast this had happened when just two months earlier I had never met a Blackburnian. My learning curve was steep, thanks to my immersion birding with Peter. And now, I maybe was going to add the Nelson's Sparrow to my birdy acquaintances. Still, I wondered if, through the many bird songs so new to me, I would hear Nelson sing.

In his *Sibley Guide to Birds*, David Allen Sibley describes ten sparrow species (out of the thirty-three that live in North America) as solitary and secretive, with secretive often modified with *very* or *rather*. Identifying sparrows requires a quick eye that takes in details in miniature: stripes on the chest or eye line or what sort of cap it is wearing. Short tail or a big bill? I found these descriptions amusing in relation to a bird not even the size of my hand. Sparrows ask for an eye focused by an adoring attention to detail. It is a lover's eye. Peter loves sparrows.

I first took stock of Peter's sparrow love when in early June he had sent me a text lamenting that Grasshopper Sparrows had not returned to the wide fields of the former dump below his house. "No Grasshoppers to be found. My heart drips a sea of Grasshopper tears." His despair over a bird with such a lively name amused me. Like the Nelson's the Grasshopper belongs to the *Ammodramus* genus (Latin for sand runner), which are smaller than other sparrows; all have big bills and flat heads. Most, Peter explained, were pathetic fliers.

Maybe, Peter wondered in a text, his Grasshopper Sparrows would never return. It was impossible not to be charmed by this man who worried over the fate of a little bird with a weak song and an even weaker flight. For now, though, we were looking for the Grasshopper's cousin. In pursuit of Nelson we abandoned standing at the water's edge and decided to find the bird by boat. We slid our long sea kayaks into Lurvey Spring Creek near Bass Harbor, which opens out onto the Atlantic Ocean. The creek spread fifty feet wide, the water running clear and shallow. Small fish—were they dace?—slalomed through the green water grass that waved in the faint current. This was so different from the Hudson River, the water there always turbid, holding its secrets. This water smiled at the world, inviting us in.

We followed Lurvey Spring upstream, against the current, away from the sound of cars clunking across a steel bridge that spanned the creek. The summer air smelled like silence and hope, that pure mixture that permeates empty northern landscapes. The stretch and pull of my paddle strokes through the delicate water felt natural, invigorating. Peter led in his white kayak, moving efficiently as if he knew where he was going. I pushed back in my seat, cradled by my boat, as I enjoyed a Catbird's whine and a Song Sparrow belting out its charge. A Turkey Vulture soared overhead. I noticed its signature wobble ("they teeter like they are drunk," Peter coached). My boat moved gracefully through the water, as the stream beckoned us inland.

After fifteen minutes of paddling, Peter raised a hand as if signaling a halt. He pointed east, toward the bank of the creek, toward a sound that could have been the grass swishing. I swung right and stopped paddling, gliding toward him until I could hear him whisper, "Nelson."

Peter sat alert, his paddles resting on the edge of his boat, his binoculars in his hands. Our boats nudged each other as we rafted up, scanning for movement, straining for another fizz of a song. I squinted into the sun, into the grasses that didn't sway or tremble, half holding my breath.

Up popped a bird, wobbling at the top of the riverside grasses. I let out a gasp, a human-sized Nelson-like song. Before I had finished my small exclamation, the bird had dropped down, straddling two stems of grass, as if doing the splits. It paused for a moment, long enough that I noted the orange racing stripe behind its eye. It darted back into the grasses.

My heart spread with happiness. We nudged the bow of our kayaks into the grasses to keep from drifting downstream. I can't say how long we hovered there, time folding up like an accordion, as we hoped for another look. All of my attention focused outward, combing the grass for movement, my ears stretching to hear another wisp of song.

"Guess that's it," Peter said, as if declaring the show over. We both settled into the seats of our kayaks as if he had said "at ease." Though I wanted more, a better look, would always want more, I was coming to see the upside of getting just a taste: we stayed hungry. So it didn't surprise me when Peter said, "Let's see what else we can find," and pointed upstream.

Nelson wasn't always called Nelson. The Nelson's Sparrow and the Saltmarsh Sparrow together made up a species known as the Sharp-tailed Sparrow. In 1995 the American Ornithological Union *split* the species, making two species from one. The two, despite this splitting, to my eyes look remarkably similar. Peter spoke with other birders of splitting species—as opposed to *lumping*—as if they might be discussing the next election. I listened amused, but still had no opinions. What did fascinate me is that two centuries into identifying birds on this continent, we were still—and perhaps always would be—understanding with finer detail the differences or similarities between species.

In the splitting of the Nelson's and the Saltmarsh one bird ends up named for a particular habitat, and the other for a nineteenth-century naturalist, Edward William Nelson. He collected (the polite word for shot) specimens of these sparrows in 1875 in the Calumet marshes near Chicago when he was only eighteen years old. Later, Nelson worked for the biological survey in the Alaska Territory. The northern climate didn't suit his constitution, so he turned to dry desert landscapes to continue his work. The desert bighorn sheep are named for him, as well as a milk snake, and ten species of rodent. In the many ways that a person can be remembered in this world I thought it would be wonderful to be forever associated with a rodent. I did not feel any loss that I would never know Edward Nelson, but I was happy to have spent a brief time with the little sparrow that carries his name.

Peter and I continued up the creek until we found a place to pull out of the water. On a grassy path a Nashville Warbler flitted in the trees.

Nashville seemed so far from these green woods, yet there was the bird, so yellow with a gray head and sporting a rust-colored beret.

Strange that someone thought it a good idea to name a bird for a place it only passes through on migration. But it wasn't just the Nashville that might mislead me, the Tennessee and the Connecticut Warblers also fall into this category of being named for places they pass through on migration, not places they winter or breed (and the Connecticut Warbler migrates through Connecticut only in fall; its northerly route passes through the Midwest). I was coming to see how at times names could confuse more than help. But I realized that confusion was only my own—the birds carried on, eating, breeding, and migrating, regardless of their names.

Further along the grassy path, our way opened onto a wide field.

"We should be seeing Bobolink here," Peter said.

Within minutes the crazy, digital-sounding song of Bobolinks played above our heads. I grinned, "What did you do, summon them?"

"It's all about habitat," Peter said, shrugging off my suggestion that he could work magic. In an open field or marsh, in dense woods or near a pond he always asked: What bird should be here? Knowing what to expect, what to look for, increased the possibility of seeing a particular species. But that meant I had to know who preferred what habitat, who lived where. Since I did not yet have that knowledge, I smiled and continued to see Peter as a magician, a bird summoner.

"That's a crazy song," I said, squinting into the late morning sun to catch sight of the Bobolinks. With their white capes on black bodies they looked like the philharmonic had taken to the air, corkscrewing about like trapeze artists.

"Think R2D2," Peter said.

That seemed about right. And yet how far from Thoreau's description of the Bobolink song:

> I hear the notes of the bobolink concealed in the top of an apple-tree behind me. . . . He is just touching the strings of his theorbo, his glassi-chord, his water organ, and one or two notes globe themselves and fall in liquid bubbles from his tuning throat. It is as if he touched his harp within a vase of liquid melody, and when he lifted it out the notes fell like bubbles from the trembling strings. Methinks they are the most liquidly sweet and melodious sounds I ever heard.

Methinks that from Thoreau's theorbo—a lute with a long neck from the sixteenth century—to Peter's robot a lot of things have changed in the world. And yet the Bobolink has not changed its song. Has not changed where it sings either. Birds are true to themselves. They aren't swayed by fashion or peer pressure, the whims of a lover. Born a Bobolink, the bird spends its life being just that, even when we humans layer on new names, describe the birds in different ways. The sureness of a bird's life reassured me that sureness was possible; the birds served as a winged model for being who you are.

How much time had I spent moving to the next house, finding a new job, leaving the next lover as I "figured out" who I was? Was I French like my mother or American like my father? Was I a rock climber, a person of movement, or a writer, a person who could stay still and stick to words? Did I love women or did I love men? I felt I had to choose, even wanted to choose in order to feel more grounded. It took me a long time to realize that the answer to these questions was, quite simply, *yes*. What I had to learn to live with was the complexity, often a back and forth, sometimes an in-between-ness.

And now, here was a new identity, birder. In those early days of birding, I didn't know how fully I had stepped into the comfortable shoes I had always wanted to wear, that I was becoming the person I wanted to be. But I sensed that something was right about all of this; I felt it in the ease of my paddle strokes as we headed downstream, aided by the gentle current, to where we had begun our journey.

Mount Katahdin loomed in the distance, a husky dome rising from a flat landscape. Peter and I were now three hours north of Mount Desert Island, north of our Nelson's Sparrow as we drove south of the mountain on a narrow two-lane road just outside of Baxter State Park. I remembered out loud the first time I had seen Katahdin. It was winter, early 1990s. A group of us, all women, had tried to summit, hauling ice axes and crampons, ropes and helmets. A vicious winter storm had blown us back, kept us from the top. After that trip, I had promised myself I would return and stand on the summit of Katahdin. But I never had. I could not think of what I had done in the past twenty-five-odd years that was more important than climb this mountain.

"You can hike up it if you want," Peter said as I finished my story of cold winter heroics and defeat.

"You don't want to?" I couldn't imagine not being drawn to the mountain. It felt like a beacon, one that pulled hikers, many of whom had spent months on the Appalachian Trail, to their final summit. Hikers through the centuries had made their way to the mountain, including Thoreau, who tried, but failed, to make it to the summit.

"There aren't any birds on the top of the mountain," Peter explained.

I nodded. I calculated that if I hiked up the mountain alone and he found some beautiful or special bird without me, I would regret that. The mountain would be there, would wait—I could come back. A bird, though, would not wait.

Peter and I drove on, navigating our way toward the Golden Road, a ninety-six-mile road that runs from Millinocket west and north to the Canadian border. The name of the road made it instantly legendary in my mind. Maybe the origin of its name came from the cost to build it, or maybe we would find gold there. The road is private, owned by the Great Northern Paper Company. That made it a logging road.

Peter's hope was that on this empty, wooded road we would see boreal species like Gray Jay, Three-toed Woodpecker, and Boreal Chickadee. I liked the sound of these birds and of a boreal forest, where it would be cool, likely filled with mosquitos, but also empty of tourists. Though I wanted to visit Baxter, I also liked that we were not doing the usual tourist thing. Birds were going to show me a world most don't see.

Once on the Golden Road, first paved then not, dust kicked up around my Subaru wagon, the road edged tightly by dense evergreen forests. A logging truck, laden with trees, now lying flat like telephone poles, approached us at a fantastic speed. I pulled over as far as I could to let it by, as a pebble pinged against the windshield.

After several miles along the flat, straight road, enough that my car was nicely dusted, I asked, "Can we get out and walk?"

Peter nodded. An opening in the woods looked like we could walk out a bit, but this was not a woods laced with hiking trails. It was a woods where little light penetrated, the fir and spruce trees growing shoulder to shoulder.

As we gathered our binoculars, a water bottle, and snacks—unsure of how far we would be able to walk—a pickup truck pulled over. A man rolled down his window. I could see a fishing pole and tackle box in the back of his truck. "You lost?" he asked.

"We're good," Peter said. He waved his binoculars. "Looking for birds."

The man nodded. "Look out for moose," he said and drove on.

I laughed. "I guess we look lost."

The smell of northern woods entered my senses, and I stood, inhaling a scent that felt like a memory from a perfect childhood.

"This is beautiful," I said, then wanted to retract my words when another logging truck rumbled past.

We walked away from the road, a narrow passage among the trees letting us in. The woods were so dense the trunks of the trees looked black, all fighting for space and for light. I stopped for a moment to listen, wished I had brought a sweatshirt. A ticking silence surrounded me, then the drilling whir of a mosquito in my ear. The path was grassy, smooth, and flat, making its undaunted way through this army of trees.

And then I saw it, a hundred yards down our path, brown on brown, only texture giving shape to the moose, a creature much bigger than I had imagined. It had a blocky head and large eyes, which stared back at me.

Peter already had his camera to his eye, the moose so close through his lens he could see strands of the stiff fur. "Nice bird," he whispered as the moose turned and, stepping deliberately, lightly, vanished miraculously into the woods.

In Thoreau's 1846 essay "Ktaadn," he describes his attempted hike to the top of the mountain. The land around him was "not lawn, nor pasture, nor mead, nor woodland, nor lea, nor arable, nor wasteland." In other words, it was a kind of raw wilderness, previously unknown to Thoreau. On his descent, "on the skirts of the clouds," he looked down on the woods where Peter and I stood. From there he notes, "It did not look as if a solitary traveler had cut so much as a walking-stick there." It still had that sense of untouched wilderness, but the logging trucks told me a different story, one that made me uneasy. Maybe the fisherman was right; we were lost.

Thoreau's untouched woods now had twelve thousand miles of logging roads. A quarter of a mile beyond the woods that framed the dirt road, the belly of this forest had been scooped out in a giant game of pickup sticks. This logging had happened faster, relative to their size, than the clearing of the Amazon. Here, in real time we were witnesses to *habitat loss and destruction.* How I wished that we had let this forest be a forest.

Peter and I would never see this leveling of the forest, only the trucks laden with trees. It wasn't that we were content not to see what we didn't want to see, it was that the narrow paths into the woods only took us so far. We only experienced the edge of the woods, still intact, and rich in bird life.

"Listen," Peter said. A clear song emerged, as if from a bigger bird. What I saw was a rusty sparrow, with a full, round chest. So handsome.

"Fox Sparrow."

The bird sat up, singing, so unlike its cousin Nelson. How different sparrows could be, I realized, some bold, some timid, some with elaborate songs, and some with just a whiff that passes for a song. Such range within one family seemed rich, complex, and worth spending time learning. I was beginning to understand Peter's fondness for sparrows.

Our passage into the woods came to a sudden end, and we turned to walk back to our car, that one bird our reward for venturing out. But that we had seen anything in such a vast landscape seemed the greater miracle.

Later that afternoon after hours of driving, walking, and searching we left the Golden Road. I felt a weight lift from my shoulders. I wanted to stop near water, to wash the dust from my face and hands. The Allagash River, stretching about a hundred feet wide, pulled us over. From the bridge, we watched as a group of teenagers slid canoes loaded high with supplies onto the river. I envied them heading off on a river trip, making their way northeast with the current. We waved them off then walked down to the banks of the river with a blanket and snacks. The clear air and quiet felt good after the start and stop of road birding.

A Belted Kingfisher streaked by letting out its characteristic cackle. On the far side of the stream, close to one hundred feet wide, it perched on the mud bank for a moment. Then it vanished into a hole.

"What happened?!" I asked.

"I've always wanted to see a Kingfisher nest," Peter said, leaning into me, his hand resting on my back. I liked that for all Peter had seen, there was still so much more. This world of birds might be infinite, one that would keep me alert, entertained forever.

We snacked on cheese and crackers, peaches not yet ripe, while the Kingfisher flew in and out of its nest, feeding its young. The Kingfisher is a dramatic bird, with an oversized head dominated by a thick black bill. Each time it launched out, executing its straight-as-an-arrow flight, it let out a rusty cackle.

I felt like we were spying, witnessing the private life of this bird. I only wished that we could see the baby Kingfishers straining toward the light, toward food, and life.

Baby birds. They were everywhere, Maine being breeding habitat for many species.

Baby Spotted Sandpipers bobbed their butts along the waterfront;

baby Common Loon rode on their mother's back;

baby Ruffed Grouse shuttled across the road;

baby Boreal Chickadees flitted in a spruce tree;

baby Black Ducks and Common Goldeneye floated buoyant on a pond;

baby Downy Woodpeckers strained out of a cavity in a tree;

baby Ovenbird and Wood Thrush scuttled through the forest.

All loveable fluff and spastic energy, these birds spoke of the fertile natural world, of the fierce desire to live. Was it possible to witness this and not think about one's own procreation? I doubt it. In the early 1800s Alexander Wilson wrote:

All nature, every living thing around me seemed cheerfully engaged in ful-filling that great command, "Multiply and replenish the earth," excepting myself. I stood like a blank in this interesting scene, like a note of discord in this universal harmony of love and self-propagation; everything I saw re-proached me as an unsocial wretch separate from the great chain of nature, and living only for myself. No endearing female regarded me as her other self, no infant called me its father, I was like a dead tree in the midst of a green forest, or like a blasted ear amidst the yellow harvest.

I could have joined Wilson in mourning a lack of children, as I too had wanted kids. The moment that I felt the tug of a child was also when

I felt the noose of age—midthirties, now or never. At the time I was living in an airy house in Tucson with my partner Sam. We tried the ways that two women try to have a child. Month after month passed. When doctors prescribed tests, followed by drugs to be swallowed and injected, I walked away.

I saw that I had channeled my maternal urges into my niece and nephew, into my students. I loved being Aunt Susan to the children of my friends. Those baby-bird-filled days in Maine might have brought on a pang of regret—after all, here I was with someone who would be a great father. But I did not join Wilson in thinking about myself as an unsocial wretch, as a childless woman. Because this moment was not about me not having something, it was about the birds having something: young, the next generation, that urgent need to reproduce fulfilled. I only felt giddy with all of the baby birds.

But I wondered about Peter as he took photos of the young birds, capturing the details of their tiny beaks and downy feathers. Was he thinking of his son? I would not know. We did not talk about ourselves; we talked about birds. They were our desire.

For our final night in Maine, I convinced Peter to give over to Baxter State Park. We were in luck, as a rustic one-room cabin on Daicey Pond was available. While I took in the view onto the lake from the cabin, Peter returned to the car to bring out our gear. Rather than carrying it down the long, narrow path, he loaded it into a borrowed canoe. I tracked him as he paddled surely toward me with duffel bags, pillows, and towels piled high. It looked like the Beverly Hillbillies on a lake.

I could see the crooked white line of his teeth, the length of his legs, the blue of his eyes (no eye ring), the white hairs at the tip of his red beard. He felt so familiar, yet watching him maneuver the clunky canoe toward shore, I reminded myself that I knew as much about him as I knew about the birds: little.

Before we left on this trip Peter had arrived at my house with flowers—daisies and some phlox—he had picked by the side of the road and a card printed with one of his photographs of a Cedar Waxwing. Inside he wrote: "I've always known you, now I'm glad to finally meet you."

Maybe all new lovers feel this way, but in this case, the sense of familiarity, so strong, puzzled me. Beyond the birds, Peter and I had little in common. I came from a world of books, grew up in a college town; he claimed he grew up in a cult, and now lived a life in the trades. His world was visual and musical. I had a tin ear. Maybe the birds themselves made us kin.

Our night in the cabin passed too quickly. The rising sun filtered through the open windows. We had been birding all day every day for ten days. I was tired. I was also happy to have learned some birds, to learn I was capable of learning, to learn that all I knew amounted to near nothing in the ocean of birdy knowledge.

Peter asked, "What do you hear?"

We were glued together in a single cot, his arms wrapped around my shoulders, my head cradled on his chest.

"Winter Wren." I paused. "Red-eyed Vireo." More. "Red-breasted Nuthatch."

I am lazeee. "Black-throated Blue."

"Right," Peter said quietly. He kissed the top of my head. "Good."

I allowed myself a small smile at Peter's praise. Maybe even heard in that *good* a hint of *I love you.* We hadn't lingered like this in the morning once on this trip. I started to form the foolish idea we would dawdle at our cabin, enjoy the lake. Maybe even go for a swim before packing up and heading home.

"Come on," Peter said, vaulting out of bed. "We're going to find a Spruce Grouse."

FLORENCE

Bighorn Mountains, Wyoming

In 1894 ornithologist and writer Florence Merriam Bailey spent a spring and summer birding on horseback in San Diego Valley, California. In her book *A-Birding on a Bronco*, she describes the joy of being immersed in migration and a breeding season in a new bird world. Every day, this native of New York State rode out on her horse, opera glass in hand, notebook at the ready. The guide she consulted was *A Manual of North American Birds*, known as "Ridgway's Manual." This 1887 bird guide, written by Robert Ridgway, weighed in at a hefty 642 pages. That season in California, Bailey "made the acquaintance" of seventy-five birds, naming fifty-six of them.

I know I would not be able to do the same if I swapped out my binoculars for an opera glass, was stripped of audio apps, and had lost my *Sibley Guide to Birds*. For this reason, I am grateful I bird in the twenty-first century. What I would like is to have all of these tools but live in the time of bird abundance Bailey experienced. But my envy gets me nowhere. So what I do is read her guides and narratives and dream of a richer bird life.

It's not just for the birds that she brings alive that I fell for Bailey's writing. All it took was opening *Birds through an Opera-Glass* and randomly landing on the *Yellow-Bird, American Goldfinch, Thistle-Bird*: "Throw yourself down among the buttercups some cloudless summer day and look up at the sky till its wondrous blueness thrills through you as an ecstasy." Her over-the-top enthusiasm is as infectious as the birds themselves.

I had started birding with the Burroughs Society the spring before and noted that there were always more women than men on the walks and appreciated that the club had an expert birder who was a woman as president. Birding seems like a gender-neutral activity—all it takes is good ears and eyes. And yet it was clear that men (Peter included) often led the walks, whether formally or informally. Bird guide companies might work toward gender equity, but most fall short. The subject of women in birding is discussed and written about in the community, and one *Audubon* article put it this way: though women make up the base, "men have the loudest voices and the most power. . . . Men hold the highest positions at the American Birding Association, the Cornell Lab of Ornithology, and the American Bird Conservancy. They dominate bookshelves, festivals, competitions, and gear and travel ads." So the fact that the Audubon organization, up and running since 1886, has never had a woman president is not in fact surprising. When I searched contemporary guides to North American birds men did, indeed, weigh down the bookshelf. And yet there, at the turn of the twentieth century, was Florence Merriam Bailey, who authored five guidebooks (one coauthored with her husband, Vernon Bailey).

Bailey also wrote several narratives of her birding experiences, including *A-Birding on a Bronco*, which takes its title from Shakespeare's *Merry Wives of Windsor*. The husband goes a-birding, leaving his wife to play. There is, alas, nothing saucy about Bailey's book, which might better be titled, "Watching Bird Nests for Hours Every Day while My Horse Grazes Nearby." Even her horse, Mountain Billy, gets bored waiting while she communes with the birds: "Ornithologists are discouraging people to wait for, and Mountain Billy got so restless under the gnat tree [a tree in which Blue-gray Gnatcatchers nest] that he had to invent a new fly-brush for himself." How hard it must be to write a book where one of the central characters—the bronco—is bored.

What held my attention in this near plotless book were the nests. Bailey details the activity around nest building, beginning with a pair of Blue-gray Gnatcatchers, who build a succession of nests—"beautiful, lichen-covered compact structures"—as each one is pillaged by other birds, most often the tiny Wrentit. The third and final nest is kept safe, and babies emerged. I admired the gnatcatcher's determination—one nest takes ten days of work, only to be abandoned—but also Bailey's patience, spending weeks as witness to this determination.

At one point Bailey sits beneath the old oak tree, watching the gnat-catchers, when she realizes there is a nest hole in the neighboring syca-more. She took delight in watching the adult California Woodpeckers (a subspecies of Acorn Woodpecker) come and go and listening to the calls of the young from deep within the trunk of the tree. When both wood-pecker adults are found dead, lying at the edge of a field, she mourned that they likely were killed by poison-laced raisins meant for the gophers on the ranch. Concerned about the young, she called over the ranchman's son, who with great effort hoisted himself up, then expanded the hole so that he could reach inside to pull out the young. Rescuing the baby woodpeckers is the most dramatic moment in the book. Bailey then raised the woodpecker babies by hand. She named them Jacob and Bairdi, and as they grew, becoming lanky teens, she had their portraits taken by a neighboring photographer. Here was a woman for whom the birds were family; Bailey's quirkiness, even eccentricity, is so natural that I knew I would have followed her into any field or down any path.

Bailey explains that her preferred method of birding was to stroll into a field, then "sit down in the grass, pull the timothy stems over my dress, make myself look as much as possible like a meadow, and keep one eye on the bobolinks." I was much too restless to become a meadow, and I wasn't even sure if meadow-ness was a good way to bird. But I did not fret about my inability to stay still like Bailey, as she had been training her whole life to be a meadow.

Bailey's young life flowed with birds and with an intimacy with the natural world. Raised in the family estate of "Homewood" near the town of Leyden in upstate New York, she spent her childhood shadowing her older brother, C. Hart Merriam, through the woods. Hart, age sixteen, made a name for himself as a member of the Hayden Geological Survey of 1871, in which he participated, in part thanks to his father, who was a member of Congress. That expedition was responsible for making Yellowstone the first national park. Hart returned from the West with over three hundred bird skins. Creatures, whether alive in the field or dead in the hand, were a part of Florence's early world. When Hart was off at medical school, fourteen-year-old Florence wrote to her adored older brother that their mother wanted to know "what you want done with the skeleton you left on the kitchen stove." So it is through her family, in particular her brother, that she entered this naturalist world that at the time was reserved for men.

Later, Hart Merriam headed the U.S. Biological Survey and was a founding member of both the American Ornithological Union (AOU) and of the National Geographic Society. In Washington, he moved in the elite circle of naturalists that included, among others, Theodore Roosevelt, before and while he was U.S. president. Florence Bailey also frequented this world, which is where she met Vernon Bailey, the chief field naturalist for the Biological Survey, who in December 1899 married the thirty-six-year-old Florence. But her status in this world was confirmed before her marriage as in 1885 she was named the first female associate member of the AOU. In 1929, aged sixty, she became a full member, along with her good friend Olive Thorne Miller (pen name of Harriet Mann Miller).

Learning about these two women—birders and writers both—confirmed for me that women were a part of this history of birds. Finding them reminded me of when I started rock climbing in the mid-1970s, and I scanned both the cliffs and the history books wanting images of women climbers. When I spied a book in my local outdoor store titled *Climb! Rock Climbing in Colorado*, I had to have it. In the black-and-white photos were images of women at the turn of the twentieth century in bonnets and skirts at high altitude. And then midbook was a three-page section titled "Women Climbers of the 1970's" that showed women moving up impressive, steep walls of rock. In there was an image of Molly Higgens stemming her way up the Green Spur, Connie Hilliard jamming up the

Bastille Crack, and then the photo that held my attention: Diana Hunter leading Wide Country, wearing a long-sleeve striped shirt, her long thin arms stretching as if out of the photo toward the next hold, her dark eyebrows arching over darker eyes. In that photo, her short hair frames her smooth face making her androgynous. But there is no playfulness in her boyishness as her eyes focus not on the next hold, but somewhere else— further out, or perhaps inward—a melancholy permeating the black-and-white photo that holds so many shades of gray. By the time the book was published, Diana had died in an unroped fall, adding to the sense of sadness that suffused the photo that meant so much to me.

Still, in 1977, I wanted to be Diana. She was a dancer, so I took up ballet lessons and did pliés to give me some of Diana's graceful footwork, which for both of us compensated for a lack of arm strength. Above all, what I felt I shared with Diana was what I saw in her eyes in that photograph—a loneliness that climbing could absorb, but not take away.

As a teen, had I seen a photo of the young Florence Merriam Bailey, it would not have converted me to bird watching. She was pretty and proper-looking, hair pulled back, clear complexion, and high cheek bones. She looked like my American grandmother, who told the story that as a girl she was taught to say "papa, potato, prunes, and prisms" before entering a room, so that her lips would gather in a more delicate, more feminine manner.

Yet looks are often deceiving, as from the start Bailey might have been a proper young lady, but she was not quiet. She agitated for change while a student at Smith College, and there she began her crusade to end the slaughter of birds for their feathers to supply the millinery industry. While in college she organized bird walks thinking that if young women appreciated the birds live in all their beauty, they wouldn't want the dead ones parading on their heads. In her efforts to transform the hearts of these young women, she invited John Burroughs to Smith to lead a bird walk. Burroughs was happy to visit: "Never has a wolf been so overwhelmed by the lambs before," he wrote, not even trying to hide his wolf-ness.

Since few people wear anything except baseball caps these days, it's hard to imagine that hats could put a dent in bird populations. But they did, and the taste for ornamental feathers brought many species to the edge of extinction. In 1885, an estimated five million birds a year died for fashion. In 1888, a report in the ornithological journal *The Auk* offered some horrifying numbers from a public sale list in London: seven to eight thousand parrots, fourteen thousand quail, grouse, and partridge, four thousand snipe and plovers. In one week, four hundred thousand hummingbirds from both North and South America were sold in one week in London, and one dealer admitted to selling two million small birds in a year. By 1903, the carnage was out of control as hats measured thirty-six inches across with plumage covering this expanse. Elevators and telephone booths—now all but extinct—couldn't accommodate these wider hats. Tables in tearooms had to be placed further apart so that women wouldn't bump into each other. Hats were serious business.

As part of her work to get people to appreciate the birds, Bailey, aged twenty-six, and ten years before she would marry her husband, Vernon Bailey, published *Birds through an Opera-Glass*. Considered by some to be the first modern field guide, Bailey described behavior as a way to identify birds. This was new for bird watchers who were used to identifying birds shot and resting in the hand.

In her guide, Bailey instructed a new enthusiast—her stated audience women and children—on how to listen to and look for birds. She encouraged people to first learn the most common birds, which I knew was great advice. There are no illustrations to her guide, only long written descriptions. Her book begins with the American Robin:

> With time to meditate when he chooses, like other sturdy, well-fed people, his reflections usually take a cheerful turn; and when he lapses into a poetical mood, as he often does at sunrise and sunset, sitting on a branch in the softened light and whispering a little song to himself, his sentiment is the wholesome every-day sort, with none of the sadness or longing of his cousins, the thrushes, but full of contented appreciation of the beautiful world he lives in.

A modern guide would condense her description to "large thrush that vocalizes at dawn and dusk, tone not thrush-like." The guide you buy today gets to the point; Bailey's guide captures the spirit of why many of us get up at dawn to hear and see the birds.

As much as I enjoyed Bailey, reading her work required another sort of patience. Those descriptions of birds go on a bit too long. And her little friends the birds are often described in such fanciful ways that it becomes distracting. There are the hummingbirds that are "pugnacious little warriors" and the Catbird that is "unmistakably a Bohemian," and though the Bluebird has a "model temper, it has not always a clear idea of the laws of meum and tuum."

And then, of course, that fancifulness can also be a lot of fun, as Bailey is filled with such affection and admiration for the birds. There's the "bronzed grackle" or "crow blackbird"—no doubt our Common Grackle—which she describes as one of our "most brilliant" beauties, one that should have the name "black opal." "He is a bird of many accomplishments. To begin with, he does not condescend to hop, like ordinary birds, but imitates the crow in his stately walk." How perfect that Bailey sees the Crow's waddle as stately.

Though Bailey's guide does a wonderful job introducing the common birds, no one starts bird watching to see the ordinary; a new birder wants to feel like she is seeing marvels beyond the Robin and the Blue Jay. We want the secrets of the world, the treasures, the birds that others pass over, ignore, or don't even know exist. And if not something special in these ways, at least something that isn't seen in the backyard, something you have to go out and find on the side of a mountain.

I was three hours into a solo thirteen-mile hike in the Bighorn Mountains of Wyoming when a bird the size of a jay flashed in front of me. My pack settled into the small of my back as I focused my binoculars. The bird flew off to a cluster of whitebark pine that grazed a high grass meadow. The trees looked misshapen with crooked limbs and trunks, sculpted by wind, ice, and snow. In these branches perched the bird who, like the tree, was not elegant or dignified. But it had confidence, a lean strength, and in its sooty plumage, it looked like a bird that worked, had a job, and was utterly no nonsense. Its dark round eyes revealed little expression, like the shiny buttons sewn onto a stuffed animal. Then the bird flew off, bounding through the air as if riding the line of a long wave. I could hear, in the distance, a grating *kraak*. A new bird, and one I would have to comb the guide to identify.

I found myself in the Bighorns of Wyoming during a residency at a writers' retreat, Ucross, located in northeast Wyoming. At Ucross I lived in a landscape I adore: flowing hills, tinted and scented by sagebrush.

The sky spread to infinity, inviting me to hike out onto the dusty land dodging rattlesnakes and talking to ghosts. In the distance brooded the Bighorn Mountains, which surely held birds I had never heard of. Yet my pleasure with my Wyoming interlude was tempered; with Peter new in my life I found it harder to commit to the near-monastic habits of the colony.

The morning of my departure from the Hudson Valley for Wyoming, Peter and I took a walk down his long driveway. He held my hand, his fingers and palms calloused from his work, intertwining with my own. The gesture made me feel at once like a teenager and an adult.

"It's been nice knowing you," he said.

"What is that supposed to mean?"

"I've known you for as long as you will be gone."

He was right. I had known him for two months, and it would be two months before I returned from Wyoming. In most birding stories that I've read it's the man saying goodbye to his wife or girlfriend as he, tinged with a bit of guilt, goes off to find his next bird. Here I was the one leaving and not for a bird but for a book. Or maybe a book and a bird or birds.

Peter wears wide solid shoes and moves as if his feet are unwanted weights. The slap-crunch of his gait punctuated the silence between us as we continued down the long driveway. When we reached the road, we noticed a spot with faint texture glinting in the asphalt. We approached the splotch in the road, which was the size of a pocket watch, four little feet splayed out from under a ridged shell.

"No," I said, as I recognized the shape of the baby snapping turtle. One more casualty in a daily slaughter that takes place on our local Hudson Valley roads: opossum, raccoon, deer, rabbit, skunk, squirrel, chipmunk. These creatures become lumps by the side of the road, heads twisted back, legs stiff. Or they become furry mats, drying with each passing day as cars rumble over, flattening the bones and guts.

Once we focused, we realized there were others. One, two, three . . . we covered a stretch of the lazy country road at the foot of Peter's drive counting twelve.

"These will all be gone within the hour." Peter made a sweeping gesture across the road. "The crows come in and clean them up." The sadness I felt over the baby turtles was tempered as I imagined the crows being fed.

When Peter and I started to see each other, the snappers were on the move toward egg-laying grounds. We found them by the side of the road, out by the train tracks, or at the edge of ponds shoveling up dirt with their hind legs to bury their leathery white eggs. If I found a turtle on a solitary walk, I would text Peter photos of the snapper, eyes wide and front legs clawing the air as it dropped eggs into the dark soil where they would incubate through the summer heat.

If the egg-laying turtles felt symbolic—our relationship flourishing in harmony with the abundance of turtle life—then the dead turtles had to be seen as the opposite, even as a bad omen.

"You think you are going to get squished on the road?" Peter teased.

"OK." Good not to project too much onto the natural world. And yet my departure day felt all wrong. Without knowing it, I was leaving on Peter's son's birthday. Forrest would have been eleven.

"I'm sorry," I apologized, even though by now I knew of Peter's direct, unsentimental but loving relationship to his son's memory and knew that he would brush off my offer to stay an extra day. He had taken the day off work to remember Forrest, would visit his son's gravesite, as he did regularly. It felt like such an intimate act, as he had built Forrest's coffin, had laid it in the ground. Peter had told me about building the coffin, staring at the measurements and not being able to cut the boards to a proper length. His grief had made what was a straightforward box into something impossible to put together. Or maybe it was that a coffin should never be so small.

Soon enough, car packed tight with books, computer and printer, bike and camping gear, I pulled out of Peter's driveway, waving and forcing a smile as I headed west. I reflected that, in order to go, the move forward, away, has to be stronger than the pull to stay. At that point the pulls were equal. I had the semester off from teaching and, anticipating that expanse of time, had a keen desire to live in the West for six weeks, to finish the book I was writing. But the pull of Peter was greater than I had anticipated.

In the last two months with Peter, the world had expanded and at the same time had focused. The birds lured us down the road and through the woods where we encountered mink by the side of the pond, or a large cat glimpsed on the driveway during a morning walk. "Bear," Peter shushed one night as we settled into bed. We tiptoed to the window to watch a mother and her two cubs loitering by the bird feeder.

I needed this bigger world of creatures and the sense of wonder that came with it. It's not that I became a child again, or wanted to become a child. It's that life in mid-age narrows; the birds had loosened that constriction. My focus shifted from my aching hands to the Song Sparrow in the bush. I didn't worry if I had enough money for retirement; I cared about what birds would still be flying when I retired. Rather than scaling back, sorting and tossing the accumulation of a lifetime, I was out accumulating new sights, more beautiful songs. I felt rich for all I had already seen and was greedy for more.

On my hike, I was nearing nine thousand feet; both the altitude and the excitement of seeing and hearing the bird made my breath come shallow and fast. A new bird! Or rather any bird at all. Five miles out from the trailhead, and I could count the birds I had seen on one hand: American Robin, the bird that travels with you everywhere in the United States; Dark-eyed Junco, a winter bird of the Hudson Valley; Red-breasted Nuthatch, with its yanking nasal call; and the Mountain Chickadee, to my eye much like the Black-capped from home. The Mountain had a black eye line like a bandit, while the friendly bird I knew well wore a large black cap that draped over its eyes.

I stood in the trail and pulled out my *Sibley's*, paging through the Gray Jay, the Pinyon Jay. When I turned the page to the Clark's Nutcracker, I was looking at my mountain bird: gray overall with a strong black bill and black wings. A white patch on the secondaries matched the white tail. Black, white, gray. It was a bird hard to misidentify. I felt a rush of excitement finding such a clear and easy match (so different from those sparrows!). Yet nothing in the picture or the description in the guidebook captured the sky pulled tight over the jagged mountains, or the dusty trail studded with boulders, or the spunk of the bird.

I waited for a while, binoculars at the ready should the bird return. I listened for that distinctive call, which Florence Merriam Bailey described as a "rattling kar'r'r, kar'r'r, as [the Nutcracker] comes flying in with strong, free flight, leading a black and white liveried band through the treetops." This sound "always stirs the blood with memories and anticipations, for he is associated with the mountain-tops."

As I waited, I savored the weight and feel of my binoculars, my hands wrapped around the solid black barrels. When the bird shot into view,

I put the eyecups to my face. Boom! There was the bird, sharp, its eye shiny like a glass gem. But it wasn't just that the bird was so clear—a combination of sharp mountain air and clean optics—that gave me pleasure; it's that with binoculars to my eyes, it was impossible for thoughts to wander; the binoculars, in limiting my view, focused both my eyes and mind. As someone who has always wrestled with scattered thoughts, to have these moments of pure focus felt like a deep relaxation. Body and mind synched, reminding me of that final pose in yoga, *Savasana*. It looks like a nap but is anything but, the whole yoga practice roiling inside of the body as if each cell has been renewed. Binoculars to my eyes, I experienced an energized calm.

Before this trip, I had bought, with Peter's coaching, the best binoculars I could afford. It seemed excessive to spend $2,000 on binoculars, but I created a simple equation: good binoculars would make me a good birder. Or a better birder. And yet I also saw that these binoculars were better than me, that they signaled to others I met in my wanderings that I should know my birds. On a local walk in the Hudson Valley, a man stopped me and asked, "That's a Parula, right?" I shrugged and felt embarrassed; I was that person driving a Maserati when my skills were more suited to moseying about in a VW bug.

Through these excellent binoculars, I savored the comings and goings of the Nutcracker before moving on. Maybe the mountain would offer up more unexpected gifts. After two hours of hiking, I stopped to lunch on a smooth boulder. From my perch, the view spread to the gray peaks towering over the Lost Twin Lakes. Those lakes were my destination, still a solid mile away. I gulped down more water, ate a cookie. The trail wound through bushes turned an orange-yellow in the early fall. Short whitebark pines held out against the ceaseless Wyoming wind. The lakes pressed the edge of timberline.

A couple with three sons had arrived at the lake before me. Mom read, seated on a rock in the sun, while Dad supervised the boys as they tossed fishing lines into the water. A rifle leaned against a rock. I puzzled over why they did not even offer a nod hello. I expected the natural world to be indifferent to me, but not people. I wandered off, wondering if I should continue to hike on to the twin of this lost lake. But my feet and legs asked for a break, so I settled on a rock, my solitude pronounced.

Through my life I have spent most of my time alone. I've lived alone, hiked alone, traveled alone. My solo existence usually felt like a contented solitude, but from time to time, a hollow loneliness took over. There by the lake something slow-moving took hold of my spirit. Maybe it was time for me to give up my long-cherished belief that I was happiest alone. Maybe I should let Peter know how much I missed him.

Across the lake, rocky scree ascended to sheer cliffs, rising over a thousand feet. New snow flecked crevices in the cliff. The thin, cold air rang in my ears. For a rare moment, the wind stilled. Dizziness from the altitude, 10,334 feet, made me drink more water. I squinted into the clear lake, a mesmerizing shade of blue, and allowed myself to absorb how inconsequential I was in such a landscape.

I thought of the Bighorns as the writer Gretel Ehrlich's mountains. In *The Solace of Open Spaces* Ehrlich retreats to the wide skies of Wyoming after the death of her love. Amidst the sagebrush and cowboy life, she becomes intimate with a landscape made for loners who love people in spite of themselves. Her perfect book had made me want to walk these mountains and breathe in this landscape that seemed wide and beautiful and dangerous enough to hold any life experience.

I had assumed that through reading Ehrlich I was prepared for the power of this land. I wasn't. The Bighorns were grittier than any mountains I had hiked before. On all of my outings except this one I had been alone, hoof prints or ATV tracks the only evidence of my species. On one hike I lost the trail, and by the time I found my bearings the sun had set. I had emerged in the parking lot shaken by all of my thoughts of what could have gone wrong. My worries were affirmed by a man camped in the parking lot, tending his horse. He seemed surprised to see me. "You don't want to be out there in the dark," he said, shaking his head. He didn't say if the dangers came in human or wild form. Or cold leading to hypothermia. Despite understanding the dangers, I hiked every weekend, the mountains both tempting and spooky.

It is this push and pull, fear and desire that I love about wilderness. And in my remote hikes and climbs, coming into contact with a power big, indifferent, and beautiful, one that is not-me, forces me to sit back on my haunches, become more fully me in order to move forward. This is the challenge that I usually relish.

But at that moment, sitting by the lake, I was craving feeling close—to the birds, to my writing, to Peter. I wanted to look at the color of an eye, or the placement of a comma, to kiss Peter's gentle lips. Instead, what I had were these mountains, shimmering white, imposing, aloof, these people who focused on their fishing poles. My urge was to pack up and head downhill, but I knew I had to rest before my six-mile hike out.

I pulled out my guidebook and paged to the Nutcracker, the bird that was now giving meaning to this hike, this day, my one great find and companion. In my short time of immersion birding with Peter I had hardly used my guidebook. Rather, I pointed to a bird and Peter named it, or I took a stab at an ID and he confirmed it (or not). I relied on him to guide me in the field. Without him at my side I had to double-check my identifications. In this case, I had no doubt my bird was a Nutcracker, and I enjoyed my uncommon certainty.

I dipped my hand in the lake to see how cold it was and then sat back, stretching my legs. I pulled out a small notebook to jot down a few details from the hike, then made my short bird list. I wondered if I hadn't seen more birds because I was too focused on my feet. The rocky trail had demanded my attention, rolling gravel that could send me flying or boulders that would make me stumble. I didn't like that looking down at my feet turned me inward, but when I had tried to look out, I tripped, slipped, or nicked my shins. Soon my gaze turned back to tracking my feet. I felt like a person who is trying to sit straight, falling into a slouch.

When I looked down, Peter's words went through my head: *Look up, look out. It's the only way you're going to see birds. Trust your feet.* He was right, that the feet find the level, that they know what to do. Peter's advice traveled with me wherever I birded during that time in Wyoming, and at the end of the day, I always called him with a report of birds seen. Sometimes if I called early I would catch him at work, up on a ladder pulling electrical wires, but most often with the time difference he was home and out on his own evening bird walk.

I got back to Ucross late after my hike and hoped he might not already be asleep. But he was awake, lying in bed.

"Reading," he said. He was reading his *Sibley's* guide as he did most nights. I wondered what he looked for as he paged through the guide, and if looking at the images of birds right before sleep made the details of bill length and juvenile plumage stick. Maybe he even dreamed of birds.

"I saw a Clark's Nutcracker," I said when he asked how I was.

I waited for him to say, "wait a minute," as he had many times before.

On this trip, out of excitement or inattentiveness, or just inexperience, I had delivered some identifications so wacky I won't admit to them. (Who knew that White Pelicans could be found in the middle of the country?!)

"Tell me again what you saw," Peter usually questioned. That was his kind way of telling me I was wrong. I liked that Peter pushed me toward greater precision as I described the shape of a bill or what a bird was doing. These details not only held the key to identification, they grounded me. I felt the demands of the birds—pay attention!—to be liberating.

With the Nutcracker Peter said, simply, "Nice."

If identifying the Nutcracker proved easy, a lot about birding Wyoming was not. Just as I was getting a handle on the birds of the Hudson Valley I had been dropped into a new habitat, a new part of the country and a new season for birding, fall, when the birds are not as brightly colored and young birds emerge, often looking nothing like the adults. The common birds that shot from the grass on my morning stroll to my studio left me puzzled. And confusing fall warblers became even more confusing in an unfamiliar landscape.

Frustration at not being able to name a bird—all birders experience this. Florence Bailey writes: "The fact of the matter is, you can identify perhaps ninety per cent of the birds you see, with an opera-glass and—patience; but when it comes to the other ten per cent, including small vireos and flycatchers, and some others that might be mentioned, you are involved in perplexities that torment your mind and make you meditate murder; for it is impossible to name all the birds without a gun." This burst of rage from the good-natured Bailey is fun to read, while at the same time I wondered at this need to name.

I realized that naming offered an intimacy with a landscape. To name is to acknowledge. To name is the first step toward love. It was like being able to stroll down the street in my hometown of Tivoli and greet every-one by name. In the unfamiliar landscape of Wyoming, the names gave me some sense of belonging as I called, hello, skittish Northern Flicker, glorious singing Western Meadowlark, valiant Black-billed Magpie. Still, so many of the birds I could not name.

"Where are you when I need you?" I would ask Peter, as I peered at a bird, phone in one hand, binoculars in the other. "OK, I have a sparrow. It has a whitish chest with a black dot at its heart."

"What else? What does the crown of its head look like?"

Sometimes my descriptions led him nowhere. "Don't focus on color. Tell me about where you are. What is the bird doing?" His questions taught me what to pay attention to. And he astonished me when, with but a few details, he could identify the Sage Sparrow (now known as the Sagebrush Sparrow), or the Townsend's Solitaire, both western birds he hadn't spent much time with. One evening returning to Ucross I pulled into the reservoir, and four odd-looking birds wandered in front of my car. I called Peter. "Goofy looking chickens," is how I described them.

"They have black bellies? Tell me more," Peter insisted, enthusiastic about the Greater Sage-Grouse. "That's a bird I want to see."

Playing "name that bird" got us around the problem that we both didn't like talking on the phone. It also ensured I never mentioned the loneliness I had experienced in the mountains.

Throughout her guide, *Handbook of Birds of the Western United States* (illustrated by Louis Agassiz Fuertes), Florence Merriam Bailey uses no possessives in bird names as we do today. She referred to the Clark's Nutcracker as the Clarke Nutcracker, adding an *e* to Clark's name. I wish her system of naming persisted because she is right, it is *not* Clark's bird.

It was Alexander Wilson who named the Clark's Nutcracker (*Nucifraga columbiana*) for William Clark, of the Lewis and Clark expedition. Clark first came across the bird in 1805 near the Salmon River, which is the headwaters of the Columbia (hence *columbiana*) and thought it a woodpecker. That made sense to me as the bird does have the undulating, athletic flight of a woodpecker. Wilson changed that identification, placing it in the crow family and naming it Clark's Crow (*Corvus columbianus*). I saw that the bird had crow-like features, like its size and boldness. Later, Audubon again changed the name to place it in the nutcracker family, the only North American nutcracker, and with a European cousin, the Spotted Nutcracker. Reading this history of the Clark's Nutcracker made me appreciate the challenges of early ornithologists trying to classify the birds. What they had to go on was observation alone.

The Nutcracker lays its eggs early in the season, in February and March, when blizzards, snow, and hail are still common in the West. They store pine seeds to feed their young through the long winter. The Nutcracker often has hundreds of stashes containing tens of thousands of seeds. They have an ability, which to many seems miraculous, to find these seeds even in deep snow. And their system works; the West is bountiful with Nutcrackers. Yet the future is not secure (of course no future is secure) for the Nutcracker because its greatest food source, the nuts of the whitebark pine, is threatened.

The Nutcracker and the whitebark pine have an agreement. The Nutcracker, in caching seeds of the tree, disperses them. Those that the Nutcracker does not eat potentially become new trees, which makes Nutcrackers the great foresters in the West. But in recent years, both the mountain pine beetle and a fungus called white pine blister rust have been killing the pines. The whitebark pine status is fragile enough that as of 2011 it became eligible for endangered species listing. So: fewer whitebark pines means fewer healthy Nutcrackers eating the fatty seeds of the tree, caching the seeds, and allowing for new whitebark pines to grow.

Scientists refer to this agreement between the whitebark pine and the Clark's Nutcracker as mutualism. *Mutualism*. It's an awful word and a great idea. Both creatures go about their lives while both get something from the other. In my mind, mutualism was at the heart of all good relationships, avian or otherwise. I knew I was getting the fatty seeds of bird knowledge from Peter. But I wondered what I offered him besides the obvious fact that I was not his ex-wife.

This idea of mutualism has been used in the research of Cornell Lab of Ornithology–trained ornithologist Taza Schaming. Schaming's writing made me want to talk to her, so I called her one afternoon once I had returned from Wyoming. I wanted to learn about the Nutcracker, but also I was intrigued to find a woman ornithologist studying this neat bird.

Schaming was warm, articulate, ready to tell me about her birds and her relationship to them. She didn't fall in love with the Nutcracker; rather she realized that the bird would allow her to live the life she wanted (mutualism!). A former ski bum, she wanted to work outside, in alpine and subalpine environments, and she wanted to engage in research that was conservation related. When she realized how little research had been done on the Nutcracker she was elated: "It's unusual to have a bird so poorly studied in North America."

She certainly got the cold she wanted. Schaming's work requires going out in winter in Wyoming (think negative twenty) with her beef suet and a bow net to capture the birds. Some days she catches all of one bird. Her patience reminded me of Florence Bailey watching nests for hours at a stretch.

At the time we spoke, Schaming had three years of radio-tracked data, much of it yet to be analyzed. Still, she already had some preliminary conclusions about the bird's habits. "The Nutcracker has three strategies," she explained. They stay put in the area where she has tagged them. They take up and move a few miles. Or, they keep moving. I laughed, and was relieved when she laughed with me. "True, it's pretty broad," she said. She quickly defended her bird and her data. "It might point to how adaptive the Clark's Nutcracker is" in terms of both diet and geography. They've been found as far east as Pennsylvania. "It's a *corvid*," she said, as if explaining everything. A corvid is opportunistic: it will eat rodents, berries, insects, and the eggs of other birds. It's the fussy eaters that we need to fret over.

"Absolutely," Schaming said, when I asked her if this was a lifetime project. (*Why do I ask such questions*, I wondered later. I could never answer such a question, and at her young age I'd resent anyone who asked me if I could imagine the future in all of its complexity, could foretell where my heart would drift.) I appreciated that Schaming didn't balk, and I liked how solidly she answered. Even if she didn't stick with the Nutcracker, she wanted to.

I admired that commitment. Only while rock climbing had I felt that I had committed myself. "Commit," my partners would call out to me as they belayed, offering encouragement. *Commit*. To the move, to the climb. The commitment had to be physical but above all mental. Hesitation could mean a fall. Climbing demands focus, and climb after climb I had to push myself to seize a hold and keep moving upward. Nothing in my life had ever asked so firmly for my attention as a sheer rock face. I'm not sure the cliffs needed me, but I needed them to learn this, how to give over, how to commit.

Now, with the birds, I saw how they also asked for my attention. The stakes were lower—no long falls threatened me if my mind drifted. Yet the greater my focus in mind and body, the more I saw, the more I learned, the richer and bigger the world became.

TWITCHING

Cove Island, Connecticut

We all have firsts that we hold close. The first time I rode a bike, five years old, down Garner Street, Becky running behind me chanting, "I'm holding on, I'm holding on," so that I believed her even when she let go and I pedaled free. The first book that captured my imagination, *The Secret Garden*, and then the first adult novel, *Great Expectations*. The first rock climb scaled at the Gunks, Horseman, a smooth face that leads to a frightening overhang. The first kiss—actually, I don't remember the first kiss. Kisses or not, love is usually involved in these firsts.

Birders care about firsts. And there are many in a birder's life. There is the first time a species of bird is seen, referred to as a life bird, or a lifer. This is the most important first, a bird that adds to the life list. Most birders can tell you the precise number of birds on their list both from the United States (which holds approximately seven hundred species) and from the world (ten thousand species). I cannot. It's not that I can't count. My failure in this numbers game is partly that I am disorganized and partly because a number doesn't tell me things about the bird, or the world, that

I am curious about; a number doesn't have the texture that opens my imagination.

After a few years of birding, adding birds to the life list doesn't come easily without travel to new parts of the country or the world. So birders keep themselves entertained with other firsts. When waking on January 1, it's important to note the first bird of the year. Maybe this carries an omen. A Robin might not be as good a year as one that starts with the hooting of a Barred Owl. My first year birding I agonized that my January 1 would start with a Tufted Titmouse.

"What then?" I asked Peter.

"Plug your ears."

What has become my favorite are the firsts of spring, as each migrating species returns or passes through toward the north to breed. Referred to as the first of the year (FOY) or first of the season (FOS), these birds announce spring. When the splintering, joyous *conkla-dee* of the Red-winged Blackbird rings over the Tivoli Bays, I know spring will soon be here, with the rush of warblers in migration. In my patch, I expect the Woodcock to *peent* in early spring, often when snow is still on the ground. And I know in mid-May to loiter in my yard at dusk, hoping that a Common Nighthawk will wing by, bat-like, toward the north.

Though I throw myself into these rituals, I remain pathetic at keeping any sort of list. I have little pads of paper scattered about the house noting in order the things I need to do or things I intend to buy. But I forget to write some things down. Or I misplace my list and so head to the grocery store without that list. Sometimes I have several versions of the same list. I had initially hoped that in birding I could learn to list and that organizational skill would filter out to all aspects of my life. I'd be that person who checks things off with efficiency. Though birding has helped me to develop certain skills, I remain a person who cannot keep a list.

"This is my first Canvasback," I said to Peter as we admired the big duck with its doorstop of a bill.

"No, you saw a Canvasback off of Esopus last May," he corrected.

I sifted through my memory deck and realized he was correct. And also that he wasn't even with me at the time.

Peter can't remember all sorts of things, like what he did for Christmas last year, or what he ate for dinner the day before. I doubt that he remembers the first time he held my hand. Yet when we walked Croton Point, he noted that here he found his first Long-eared Owl, American Coot, and Rough-legged Hawk. Admittedly these are not everyday birds, but he can also tell me where he saw his first Great Crested Flycatcher or Black-throated Green Warbler.

Certain bird firsts are etched in memory. Sometimes it's the bird itself, so rare or spectacular in color that how could I forget? But often a first is vividly remembered for the place or the person I was with. My friend Jody showed me my first Northern Gannet plummeting into the Atlantic Ocean off of Cape Cod after we had walked across the Provincetown dunes. "You can't see it from here, but they have blue legs," she told me, making these daring, athletic birds that much more wondrous. The Vermilion Flycatcher—too bright a red not to remember—that I spotted in the desert southwest on my first organized bird walk remains distinct. "You have a good eye," the guide said, fueling my sense of excitement at spotting something so very obvious, a deep red bird in an uncolorful landscape. These are memorable birds, but I did not know when more plentiful birds like the House Wren or Eastern Phoebe, the Chipping Sparrow or the Wood Thrush came into my life. I couldn't say, and that they don't have an arrival date in my life marked in my guidebook makes me feel negligent, like the parent who favors one child over another.

On a weekend in November, I embarked on yet another birding first, a trip just to see a bird. Not several birds, but a single bird, a rare bird. Birders call this twitching. Peter and I twitched a Fork-tailed Flycatcher that had arrived on the coast of southwest Connecticut.

The Fork-tailed Flycatcher is a bird that belongs in Central or South America. For some reason this bird turned north instead of south in its journey that year. Perhaps he had mistaken our fall for spring. It's a common bird in its own territory, but here in the northern United States it has landed only about one hundred times. Someone reported this bird on a Wednesday in early November, at a nature preserve near Stamford. Peter told me about the bird's arrival and for several days tracked that it was still there. He made sounds of wanting to drive down and get his life Fork-tailed Flycatcher.

I didn't want to get in a car to drive anywhere. I had recently returned from Wyoming, full of words and an almost completed book, rattlesnakes and new friends, smells of the Bighorn Mountains and the sound of elk bugling. After three days of driving cross-country it was Peter's driveway I pulled into. He stood on his deck, holding a large bouquet of flowers. He had cleaned his house, clearing out a small space in his closet for my clothes.

I have shared a home with another person only once before, when my girlfriend Sam and I bought a house together in Tucson. Sam and I shared the roof repairs and the plumbing bills, discussed what color to paint the living room walls. The wooden cabin perched on top of a hill outside of Woodstock though, was not going to become a joint project. It was Peter's house. In the bedroom, his ex had painted swirls of color across the ceiling, and the bed consisted of two singles (firm for him, soft for her) squashed together. I referred to the cavern running between them as the Grand Canyon. Photographs of his son, bald and smiling, lined the walls, stuck to the fridge, or dangled from the lights. When Peter was out of the house, working, I would often stare at the photos, wonder at the boy's beauty, wishing I had known him, wishing he were there with us now.

The house I walked into on my return from Ucross was monkishly tidy. The neat wood stack I had noticed on my first visit was an indicator of Peter's meticulousness in life. In this context, I became aware that I am not quick to fold clothes and put them away, and where my shoes come off is where they stay. I have no problem falling asleep with a mess in the kitchen sink; dishes get done in the morning while I make coffee. Some days I tried to adapt, imagining we lived on a boat or in an RV and everything had to be shipshape. In the end, my efforts always made me feel like a fraud. Maybe I could become that organized person, but soon I realized I didn't want to.

This is what it means to couple mid-age, to take on the elaborate history of another, to have more years past than future. You see yourself. I accepted what I saw. And with that came a relief, as my desire to bend like a contortionist to suit another person vanished. Or maybe I just wasn't so bendy anymore, my body and spirit stiffer.

What this meant on a practical level is that I didn't attempt to redecorate or change the landscape of the kitchen; I didn't try to make his house our home. I did small things to make the house comfortable like gather kindling for the woodstove, run loads of laundry, or make soups that simmered on the stove for hours. Yet with my six pairs of underpants bunched in a corner of his closet, this was like camping out, living in an awkward in-between place. That was fine because we both knew that this togetherness would last until January, when I returned to teaching and my own house.

Living with Peter had many perks, beginning with the fact he delivered coffee in bed (a first for me). We started the day with a walk down the driveway to the swamp where six months before he had shown me the Red-headed Woodpecker. Every morning became a birding refresher as I paid attention to details I had not before noticed in familiar birds, like the striking white eyeline of the Carolina Wren, or the fact that the Northern Flicker I had met in Wyoming was red-shafted, while these at home were yellow-shafted. And as I had learned in Wyoming, birds in fall looked different than in spring, colors muted or stripes added or subtracted. Back home I played the bird songs on my iPod hoping to graft them onto my heart. Would anything stick? The songs, the color of the outer tail feathers? The fleeting nature of it all unnerved me. In these moments of doubt, I would remind myself of that long look at the Red-headed Woodpecker. Long: perhaps one minute. Maybe one minute is enough. Maybe what the birds would teach me is you can't hold onto anything, not a bird, not a person. It seemed, at middle age, a good lesson to learn.

Peter also taught me the art of the weekend. Because he worked for himself, he never gave himself much time off. Monday to Friday he was at a job by eight and home at four to shower, return phone calls (that rare electrician), then take a walk down the driveway. At the end of this regimented week, we went out to dinner, always at the same Mexican restaurant (where Peter ordered the same chicken dish). We never had friends, mine or his, join us for these meals, as we existed in our new-love bird bubble. I was and was not OK with this. I did not want to be that friend who vanished when she found a new love. But I wasn't sure if Peter or the birds were to blame for the fact that I returned even fewer phone calls.

Saturdays and Sundays were for the birds. Sometimes we joined an organized Burroughs club walk, those that Peter had signed on to lead. It was usually a mixed bunch of excellent birders and people who had nothing else to do on a Saturday morning. It was a different experience to move about as a group, sometimes more eyes in the field bringing in more birds, but more often our size, movement, and the inevitable catch-up conversations scared off the birds. "We're like a herd of elephants," Peter complained. Or sometimes he was more direct, "Everyone shut up."

In the past, my weekends had blurred with the week, a continuous cycle of writing, grading papers, answering emails, and tending to house business. This fall, free of teaching, I let go of the writing for two days, only returning to my pages on Monday mornings. The time off was refreshing. What I soon discovered is that coming home and writing down our birding outings kept both the outing and the birds near. I started to fill wide journals with the details of sunrises and rain showers, of bird colors and song. But also hiccups of hurt feelings or moments of pure joy.

Our biggest decision every Saturday morning was choosing where to bird. Locally we had many options—farmland or woods, marshes or open water. The Hudson Valley is a rich and diverse and beautiful place for birds (and people). Where we chose to bird taught me more about migration and habitat, what birds might be where and when. That late fall we spent a lot of time in farmland looking for sparrows migrating south.

On this Saturday morning, not long returned from the West and just settling into my new routine with Peter, I decided that though I didn't look forward to the drive to Connecticut, I did want to see the Fork-tailed Flycatcher. More: I wanted to be able to say I had *twitched*, a word I found bewitching with possibility.

There's a story to the origin of the word "twitching," related by Mark Cocker in his brilliant and funny *Birders: Tales of a Tribe* (which should be titled *The Best of British Twitching*). Two of the top British birders in the late 1950s used to bustle off to find birds on a Matchless 350, a motorcycle developed for use in World War II that resembles something just this side of a moped. They brought with them Jan, a "back-warming, lap warming dog." Despite the extra doggy warmth, the two would arrive after long trips often wet, frozen, and trembling. This is Britain, after all. Soon, going for a bird was referred to as "being on a twitch." Which sort of makes sense, as twitching is defined as jerking, plucking, hipping, convulsive or spasmodic movement. There you have it, the word twitching, now accepted around the world. The *Oxford English Dictionary* defines a twitcher as a "bird watcher whose main aim is to collect sightings of rare birds." Twitching: "obsessive or enthusiastic bird-watching for rarities."

Because stories of twitching often include long cold motorcycle rides or days standing in beating sun, squinting into the distance, and they almost always involve thirst, hunger, and exhaustion, twitching can be seen as heroic, and twitchers the charmed, long-suffering explorers of the birding world. And yet Cocker writes, some deem it to be "fanatical, self-indulgent, thoughtless, uncaring, competitive, frivolous and, despite being rooted in an obsession with natural history, as actually anti-environmental." I held all of these negative impressions, especially the last one. Since I started birding, my carbon footprint had expanded in ways that upset me.

Here's the thing about taking the moral high ground. I could stay home and feel good about my impact on the planet, but then I'd miss out not just on seeing this bird, but on the whole experience of twitching. I ached at the thought of missing out on a great adventure. The Fork-tailed Flycatcher posed a moral dilemma.

Peter, who all week had seemed fascinated by the Flycatcher, remained oddly noncommittal about chasing the bird. We spent the morning birding near home, scoping the windy Hudson River to get long-distance views of Red-breasted and Common Mergansers. I was piecing together the birding that satisfied me and the birding that left me indifferent. Driving to a location where I peered into a scope at a fleck bobbing in gray-green water fell into the indifferent category. Surely the Flycatcher would be more exciting; I pushed again for the drive to Connecticut.

And won. At 3:30 in the afternoon, after a two-and-a-half-hour drive south then east through the relentless traffic of the northeast corridor, we pulled into the Cove Island Wildlife Sanctuary. The sanctuary consisted of a few dozen acres of former landfill situated at the southern end of a wide parking lot that faced Cove Harbor, which led out to the Long Island Sound. A groomed park in the distance encouraged dog walkers and joggers; an ice rink flanked the western end of the lot. A bunch of cars with Audubon decals indicated where to park. What had been a muted *why not* excursion transformed into eagerness as we raced toward the trailhead. At the entrance to the park, three volunteers mingled, talking and joking with each other.

A volunteer greeted us with open palms, signaling a brief halt, to steady us or slow us down. "The bird is here," he said, reassuring us. I noticed a chalkboard leaned against a small wooden shed with five points:

1. He's still here.
2. Stay on Trail.
3. Bathroom @ Ice Rink
4. Take Brochure
5. Send money

The volunteer pointed to the far end of a wood-chipped path where we could see a huddle of humans. "Stay on the trail, please," he called after us. It was an easy, flat walk for fifty yards or so on a wide trail. We stood a few feet from the group. Some flicked through images on their cameras, some peered through scopes, others stared at a feeder that hung from a tree across an open, grassy area. There was a disconnected, idle feeling.

Peter tapped my shoulder and pointed toward a tree. There, perched on a branch, sat the Fork-tailed Flycatcher. No adventurous trek through marshlands, no tromping for miles through woods, no hours spent holding my breath lying on my belly until every muscle ached. Easy twitching. Disappointment hovered in the air—I had wanted to work for this bird; I wanted a story.

In the absence of a chase was the bird itself. It had a white, compact body that shone in the sun. Hanging loose was its long dark tail, like a scarf thrown on for dramatic effect. In flight, its tail fanned out to reveal its impressive forked-ness.

A tanned woman named Gail, who had a disheveled, warm smile, gestured toward her scope. "Go ahead," she offered. "I've been here for the past three days." Three days she had stood on the wood-chip trail transfixed by one bird. Next to her stood her husband Tom, "one of the best birders in New York," Peter whispered to me. I liked them, their easy generosity in sharing the scope, their pleasure in the bird.

The scope brought the bird in close, its little body upright, its long tail draping a bit unwieldy. If you cut off its tail the bird appears like an Eastern Kingbird, a bird classified in the same *Tyrannus* genus as the Flycatcher. From time to time the bird dropped, sallying out to snag a bug. The broad beak of flycatchers is designed to snatch bugs midair, which this bird did with precision. In flight, that long tail made sense; the bird was an acrobat, looping through the air as if riding a great fast Ferris wheel in the sky. It soon returned to its tree limb where it opened its beak wide, chucking up the carapace as well as other tough, undesirable bits of the bug.

"Ooooh," the crowd sighed, with each sudden flight and snag.

A crowd of about twenty milled about, chatting like old friends who had already covered all of the important subjects of life and were now commenting on a good meal. "Beautiful bird." "Amazing bird." These words of praise fluttered into the air, half spoken to the sky and half to no one. One woman had flown in from Las Vegas just to see the Flycatcher, while a man had driven down from Vermont. Most, though, had shuttled in from New York City, thirty miles away, or had made the trip from within state.

Who were these twitchers? Though Peter and I birded with our local club, this was the first time I had met such a range of birders. The way they spoke to each other, it seemed a few of them knew each other, and together formed a loose tribe of expert and fanatic birders. I imagined a life where Peter and I got to know them, looked forward to seeing them when the next special bird flew in.

I asked aloud to no one in particular: What is going to happen to this bird? The bird, oblivious of all of its admirers, also looked unaware it had landed in the wrong place, where flies would soon vanish and snow would soon fly. It simply carried on its bird life, eating, pooping, perching.

Tom shook his head, "It doesn't look good."

The Flycatcher probably wouldn't find its way south. This realization muted my enjoyment at seeing the bird.

A Cooper's Hawk, all streamlined power, winged its way through the small clearing, drawing our attention away from the princess Flycatcher. A gasp rose from the crowd. What if the Coop had taken the Flycatcher? someone joked, as several let out a nervous laugh. The Flycatcher continued on it way, unperturbed by the murderer in its neighborhood.

The volunteers who had created and maintained this park milled among us, cautioning those who stepped off the trail not to squash the native plantings. Over one hundred people had visited this bird the previous day, and the rush of people had forced them to rope off certain parts of the park in hopes of protecting shrubs and flowers no longer in bloom. The roped area also kept overeager birders and photographers from trying to approach the bird too closely. This meant that we twitchers all bunched together, out of the flight path of the Flycatcher, who needed some space as it winged out from its perch. From time to time it did a lap around the open area, passing a set of feeders cloaked with local birds. I wondered if the House Finches, Black-capped Chickadees, and White-breasted Nuthatches even noticed this celebrity in their midst.

Mostly, the bird swooped, snagged a bug, then returned to its favorite branch. The self-possessed elegance of the bird continued to dazzle. After an hour and a half of eavesdropping while watching the bird, I felt satisfied, ready to leave. I did not have Gail's stamina to spend three days with this bird.

Before we left the sanctuary, Peter flicked through his photographs, all disappointingly dark in the low light.

"Let's find a hotel, come back tomorrow, and hope the bird is still here and the light better," I suggested.

We drifted through an odd evening eating a meal foraged from the salad and prepared foods bar at Whole Foods and sleeping in a sterile hotel room near Interstate 95. The next morning Peter could not find his bins. We flipped over pillows and looked under the beds then rummaged through the car.

"I must have left them in Whole Foods," he concluded.

"You what?" I asked, unhelpfully. If this was the adventure of the trip, it was a lousy one, more an unfinished morality tale than an inspiring saga of fortitude and courage.

On one level I appreciated that I wasn't the one causing the morning detour and delay; most mornings Peter sat in the car calling out, "Let's go, let's go, let's go," while I scrambled for water bottles or tried to locate my eyeglasses. Not being on the hook for this one, though, was not much of a pleasure. Peter was a mess until a sleepy Whole Foods employee unlocked the sliding glass door and handed over the binoculars.

Back at the park, the Flycatcher rested in the exact spot where we had left him the night before. At 7:30 in the morning, a group had already gathered, including a woman in a huge parka who had flown in from Colorado.

"I've got to stop the listing," she announced. She had 660 birds. "Listing is taking away from the fun of birding," she claimed. I couldn't disagree with her. I saw that the most aggressive listers in the crowd—and they announced themselves by declaring what number this bird was for them—seemed to take up the sort of space the high school bully occupied.

The bird warmed itself in the morning sun, the white of its chest brilliant. As it preened nonchalantly, then hawked for a bug, I started to feel bad for the Flycatcher. It couldn't snag a bug or poop without people commenting, or photographing the event. It was like a newborn, each movement adored. It was like a movie star, each movement tracked.

A fortress of photographers—the brotherhood of the long lens—all stood pressed close to the cord that kept them from moving further down the path. With their tripods spread wide, they took up the choicest spot. I was grateful Peter was not a member of this fraternity, as he always carried his camera, rather than setting up shop like these guys who swapped details about lenses and settings. Peter claimed he knew nothing about his camera. "I put it on automatic and push the button," he said to anyone who asked. I knew that couldn't be true because his photos were sharp, vivid. His disclaimer, I came to understand, kept other photographers away.

"You boys are taking up some real estate," I joked.

No one turned to acknowledge me or to note my admittedly passive-aggressive comment.

Bird photographers. I'd heard others complain about them: obsessive; obnoxious; arrogant. They all had one goal: a good shot. Many were willing to do a lot to get that shot, including get too close to a bird. In the process, they might scare off a bird or obstruct the view for others. In this case, these guys had the best view blocked off.

What did the bird make of these long lenses? Did they challenge its already challenged life? From the start of bird photography, ornithologists were aware that the desire to capture intimate photographs of birds endangered birds. Frank Chapman, writing in the early part of the twentieth century, recalled being in the Farallones, off the coast of San Francisco, when, trying to get a good photograph of Brandt's Cormorants, he scared them off their nests. Western Gulls swooped in and ate the eggs. In Florida, Fish Crows looted the nests of Anhingas that Chapman had frightened.

So though photographers don't shoot to kill, the result of being an overeager photographer can be death for a bird. I don't know how the photographers might have hurt this Fork-tailed Flycatcher, but they sure had altered the atmosphere among the bird admirers gathered at the Cove

Island Wildlife Sanctuary. Everyone appeared tense, a bit competitive in an odd way as people bragged of their year lists, life lists, of places they had been and birds they had seen.

One of the volunteers, a sincere-looking middle-aged woman, kept asking people to stay on the paths. For a while we all obeyed. Then a photographer just couldn't help himself; like a disobedient child, he strode out, tromping through the grasses. A few people *tish-tished*, some yelled at him, while a chorus grumbled displeasure.

I shuffled on the path waiting for the bird to come near as it pursued its morning meal. When it flitted past, the click of camera lenses sounded like muted machine guns. The tiresome loop of discussion about lists and what a great addition this bird made to that list washed over me like a blanket of boredom. These birders seemed to have lost their sense of humor, had forgotten listing is a game.

I come from a game-playing family, so I understand the passion of a good game. As I observed my fellow birders, I thought of the card games we played as a family, sitting around the table dealing a hand of Hearts or Oh Hell, or setting up for a rapid-fire game of Pounce. We laughed and yelled, and emotions sometimes got messy as a competitiveness washed through us. No one held onto these emotions, though, laughing them off by the next morning as the score sheet left on the kitchen table took on coffee stains, became crumpled, and was eventually tossed away. So I couldn't criticize these listers for the game they played, but I wished that they had remembered that games are supposed to be fun.

"Let's go," I said, hoping Peter would agree we had reached the limit of tolerance. He had.

As we walked down the path, back toward the kiosk where the volunteers still stood welcoming birders, a raucous cackling passed overhead, then a flash of green-yellow. Monk Parakeets! The Monk Parakeet, often kept as a pet, belongs, like the Fork-tailed Flycatcher, in South America. Since the 1960s, Monk Parakeets, released or escaped, have adjusted to their wild lives in the Northern Hemisphere. For me the surprise of seeing the Parakeets rivaled the pleasure of visiting the Flycatcher.

We waved goodbye to and thanked the volunteers, and I slipped twenty bucks into their bucket to help repair the plants damaged by the photographers.

As we neared the edge of the park boundary, one hundred feet or so from the kiosk, Peter whispered, "I wonder if we can't get into the northern end of this park." I realized that Peter, like me, just wanted to get away from the crowd, not the bird. It would be great to see the bird in silence, on our own.

Once out of sight of the volunteers, we turned into the trees, following a trail that ran along the fence line on the border of the park. The world stilled as if after a great wind. The tree limbs, now bare in the late fall, draped over the chain-link fence. A smell of ocean, salt and maybe dead fish, drifted in, reminding me that Long Island Sound was right there, across the wide parking lot. With the smell came sounds, Goldfinches making their potato-chip call as they bounced overhead and a House Wren playing its musical medley into the trees. This felt like birding, not what we had just been doing, which felt like a casual and not entirely welcome brunch. I turned to share my new settled happiness with Peter, who wasn't looking at me or the sky or his feet but rather straight ahead, his eyes focused as if by focusing he could keep something from falling over, from flying away. He nodded, the faintest nod, the bill of his baseball cap directing my gaze forward and a little up. Thirty feet in front of us rested the bird.

How astonishing that the bird had liberated itself from its fan club and had moved its business from the southern end of the park to this clump of trees at the north end. Maybe the Flycatcher was sick of the crowd as well.

Peter stepped to the side, squatting near the trunk of a tree. He put his camera to his eye as the bird lit into the air, then plunged to snatch a bug. Then, as if creating a circle, it cartwheeled back to its perch. It was exquisite to watch, the flurry of such long tail feathers and the arc of the long wings suggesting a bird much bigger than ten inches long. It then returned to a pint-sized bird as it perched on a limb. Each time, the bird appeared pleased with himself. I certainly would have been self-satisfied pulling off such stunts with ease.

In our secluded spot there was no swapping information about cameras or bins or life lists. Alone with the bird I captured a feeling of discovery. I imagined what a thrill it would have been to be the first to find this lost or wandering bird from South America. If twitching didn't feel like real birding, I now knew that being the one who found the bird others twitched would be a thrill.

I wondered about who had found this magnificent and stray bird and if she was among the crowd there, having her fifteen minutes of fame. Once home, it didn't take me long to track down Tina Green.

Like most good stories, hers begins in the routine of the daily: she was having her car repaired at the Chevy dealership in Darien. Not wanting to wait, she asked them to drop her off, with her scope and binoculars, three miles down the road at the Cove. "It worked out really well," she said, as if speaking of an event from the day before. "The Flycatcher was one of the first birds I saw. It was a beautiful, sunny day." I snugged into my chair at my desk at home, the phone pressed to my ear as I sensed with pleasure that she might tell me step by step what happened. "When you walk there, you go by the kiosk. When I got to the intersection, I looked over to the right. The sun was shining brightly from the east. I saw this bird. The white stood out. I knew it was something different right away, even without binoculars. *What the heck is that?* I picked up my binoculars. *Oh my God.*"

"Finding a rarity is that throat-drying, breath-snatching, chest-clutching, stomach-churning, heart-thumping, hand-trembling moment we all dream about. In birding terms it has no equal," writes Mark Cocker in *Birders*. Listening to Tina made me believe this is true. I wanted her story to go on, to hear of a victory dance. What followed proved most ordinary. She made phone calls, put her find on the Connecticut birding Listserv.

"Other people got to see it. This is what is most important."

I almost laughed. Birders are fanatical this way: you *must* share what you find. To keep a good bird to yourself will ensure you will spend the rest of your life in that special, no-doubt birdless hell made only for selfish birders. The upside of this is that it cultivates a generosity of spirit among birders, which I admire. This generosity of spirit counterbalances the fact that many birders are serious penny-pinchers, or even flat-out cheap, as they live on a shoestring in order to take off and bird.

Tina stayed for a while that morning with the Fork-tailed, then visited later in the day with a friend. She didn't visit daily and was not "a part of the whole spectacle."

"Had you ever found such a rare bird before?" I sensed that finding one rarity would produce a hunger for more, a particular ornithological greed taking over. Still, I expected her to say no.

"Uh, yes," Tina Green responded.

This is when I learned that one rarity is not like another. Or, put another way, not all firsts are equal, which is an obvious truth.

In 2009, after just a little over a year of birding, Tina found a Western Meadowlark in Connecticut. It's a "fourth state record," so a very rare bird for the state, she explained, while the Fork-tailed is an expected vagrant or overshoot along the coast.

On that day, four Eastern Meadowlarks decorated the field, along with one bird that looked different. None of them were vocalizing. Unsure what the bird was, yet sure it was different, she posted her sighting to the Connecticut Listserv. "I didn't know the etiquette of posting and stuff," she admitted to me. Birding etiquette! I knew the bird world had its own history, ethics, heroes and heroines, but now—manners! I had some sense of that from Peter who groaned over some posts, people making wrong identifications, clogging the list with ordinary birds or stupid questions. And—one wrong ID and it would take years before people trusted you again. It seemed that entire reputations in the bird world could be made or destroyed by hitting the send button. I felt lucky that Peter served as my Ann Landers coaching me on the subtle ins and outs of the etiquette of bird posts.

"I had no idea how rare it was. The Listserv went crazy. It was a Western. Someone got a recording that night or the next day to be 100 percent sure. It was such a difficult ID."

Finding a hard-to-ID rarity or an easy-to-ID vagrant, it was clear that Tina was most proud of the former. She's the one who picks through the birds, who pays attention to differences. She started giving me insight into how she had found her rarities, perhaps imagining I had called because I was hungry for my own rare bird. Was I? Not really. Not yet. I still felt like every bird was a rarity and wanted to hold onto that sense of the specialness of every bird. Yet Mark Cocker writes that if you are "not bothered with finding rarities you can't really be a birder."

Would I have the skill and the confidence to find and ID a rare bird? Or rather, when would I have the skill and confidence to find and ID a rare bird?

"Go out in the field. Study field guides at night. Look at field marks of birds you might not see," Tina coached. "If you happen on one, you'll say 'oh, that's different.'"

"Any others?" I asked, curious how many rarities we are each granted in a lifetime.

For the past three years, Tina had been on the alert for a Tufted Duck in Connecticut. A Tufted Duck is the Eurasian counterpart to our Ring-necked Duck, a bird that from time to time floats its way to both the East and West Coasts. "I wanted one here," she said, admitting that her state list is important. She knew the Tufted had to be out there. "I bet people don't spend enough time looking for it among the scaup," she reasoned, referring to the Greater and Lesser Scaup. From the shoreline, these three species can look similar, just black-and-white ducks bobbing in the waves. So she devoted her time to picking through the ocean ducks, "intent on it, specifically." She found one. "It's a combination of luck—right place and right time—and being focused on it." *Commit!*

Knowledge, focus, chance—but surely some birders are luckier than others. And surely those who found rare birds took it as a sign that they were special, even blessed. Did that mean that if you did not find special birds, you were not blessed? No, it only meant that you were not birding enough. I set out to bird with even greater focus, greater devotion.

CHRISTMAS BIRD COUNT

Ulster County, New York

In 1973, the writer, editor, and adventurer George Plimpton, who ac-
knowledged that his birding skills were sketchy, joined his first ever
Christmas Bird Count in Freeport, Texas. His sector leader was Victor
Emanuel, a man now famous for his first-class bird tours, and one of
Plimpton's good friends. Plimpton describes, in an article for *Audubon*,
how Emanuel rallied his birders:

> All right, let's try to get both cormorants, the double-crested and the oli-
> vaceous. Get close. Compare. It's the only way. The green heron is a prob-
> lem bird, and so is the yellow crowned night heron. And the least bittern,
> a tremendous problem! We've only had it once. Flush him out. He lurks in
> the cattail areas. Leap up and down. Clap your hands. That sort of behav-
> ior will get him up. Ross' goose? I'm concerned. We only had four last year.
> Look in the sky every once in a while for the ibis soaring. Search among the
> green-winged teal for the cinnamon.

Emanuel sounds like he's delivering a pep rally combined with a plan for a military maneuver. The day promises excitement and to be a bit daunting, as any endeavor is that others take too seriously. Before my first Christmas Bird Count (referred to by birders simply as CBC) in December 2010, I puzzled over this event that Peter spoke of with a hushed reverence as the real Christmas, the holiest of days.

"We're going to count birds?" I wondered aloud, curious that I had to remind Peter that birds move, they fly. "This can't be considered an exact science." Here I was, the impressionist of life, faulting others for their imprecision.

"The thing is, if I find twenty-five robins one year, the next I'll find twenty-four, and the year after twenty-six," Peter explained. "It's not the data from one year that matters, it's the data over time that matters." He sounded like an advertisement sent straight from Audubon central.

Fortunately, there are years of data. The Audubon Count has a long history, beginning in 1900 when Frank Chapman came up with the idea of counting birds as a substitute for the "side hunt." This tradition involved hunters going out on Christmas day and killing anything in fur or feathers that crossed their paths. Teams competed for the most killed creatures. There's a photograph from one of these side hunts of a dozen or more men standing on and around a wooden wagon, coyotes strung up by a hind paw, their bodies draping down, limp in death. How grateful I am that we've abandoned this killing tradition.

Chapman announced the first Christmas census in *Bird-Lore*, the journal that he founded and edited, and that later became *Audubon* magazine. It would have been easy to miss his call to count as it is wedged between a photo captioned "Guess this bird" and a cheerful poem titled "The Rev. Mr. Chickadee, D. D." whose sermon consists of the advice to be happy, be diligent and brave. This original bird census had few rules, just the mission to go out and see birds, then send the results into the journal as soon as possible. Laced into these few guidelines is a sense of wholesome competition of who might find the most birds, or the most unexpected birds.

For Chapman's first count twenty-seven people combed twenty-five random locations. It was a warm Christmas Day in the Northeast and no one saw either Red or White-winged Crossbills. Clarence Brook from

Keene, N.H., counted for three hours, noting one Northern Shrike, one Crow, and sixteen Black-capped Chickadees. Many started searching for birds around 8:00 or 9:00 in the morning, with only one person out there at 6:30 a.m., in Oberlin, Ohio, who in three and a half hours found fourteen species.

All of this sounds quaint, and the numbers tiny compared to what unfolds today with the Audubon organization overseeing more than two thousand counts in the Western Hemisphere. That means thousands of people combing swamps, forests, fields, and shorelines for birds. No longer is it held only on Christmas Day, but rather on a day between December 14 and January 3, which takes into account work and family obligations and also allows for birders to participate in more than one count. There are now a lot more requirements and rules. Eight hours or more in the field is preferred, not just a few random hours, and predawn owling is encouraged. The area to be counted is within a circle with a fifteen-mile diameter. That circle is then divided up into manageable areas that two counters can cover in a long day, which means that most circles have over two dozen participants. What remains, connecting the early twentieth century to the twenty-first century, is that sense of competition, now not just wholesome but vigorous. I doubt Chapman had any sense how big his idea would become. Yet how marvelous that this man, who had such a near-religious conversion to the birds, launched this holiday that for many birders is the most sacred day of the year.

Though the whole thing sounded tedious—a long day mostly spent in the car, not *birding* but counting birds—I anticipated my first CBC as if it might be my induction into the holy world of birding. I prepared in the only way I knew how for a celebration: I cooked. When I slipped into the car at four in the morning of December 18, I packed turkey soup, an onion quiche, ham and cheese sandwiches, blue corn chips, carrot sticks, homemade biscotti, and a thermos of tea. Whatever else might happen, we would not be hungry.

We drove west on 199, the town of Stone Ridge still asleep as we sped through toward our sector. Once we pulled into our territory, Peter stopped so that I could note time and mileage of our nighttime hours. And then we were off, Peter slaloming the winding roads of our sector, the headlights of the car slicing a narrow path through the darkness and trees.

A disorientation settled in as Peter turned left, then right, while I marveled over the size of our sector. After twenty minutes, we jounced down a dirt-packed road, then crossed a small creek, before Peter pulled over next to a wooded hillside.

We slid out of the warm car into fifteen degrees of cold. A frozen silence enveloped the woods, as the engine ticked from warm to cold. I stared, open-mouthed, toward the heavens, a canopy of stars lighting the swampy area in front of us and the dense woods that formed a wall of darkness. I tried to see into the darkness, the open space that was the road leading into denser woods, the clearing in front of us framed black by the night. I shivered, not at the cold but at the beautiful and overwhelming limitlessness of the world. And somewhere out there might be Saw-whet Owls. With my first icy breaths, I felt that to be absurd. Nothing was out there, the whole earth frozen.

In the weeks before the count Peter had scouted this sector, which had been his to count for the past four years. He drove around to get a sense of what birds were where, which houses might have a feeder up luring in birds, and what habitats he might have missed in past years. During that reconnoiter, he had realized that this remote section, dense with conifers, was perfect territory for Saw-whet Owls. All owls are special, seen as powerful, mysterious, and wise, but the Saw-whet is so tiny—only three ounces of bird—it's all of those things as well as adorable. With a proportionally big head and luminous eyes it has a cat-like aspect. The Saw-whet is more silent in the night than its much bigger Barred and Great Horned cousins. So finding one by luck or chance as it lets out its song is unlikely.

Peter had never heard or seen any Saw-whets on CBC before. "That doesn't mean they aren't here," he said. "People don't see Saw-whets, because they don't look for them."

Peter paused to be sure I was following his logic, while I mused over whether this was true for all of the best things in life: to find the good bird, you have to look for the good bird.

"You have to believe in them," he said, as if asking me to believe in God.

God appearing on this road seemed more likely than a Saw-whet, though both asked for me to imagine, believe, hope. But the bird also asked me to be logical: think habitat and what is possible. Stretch for what might be out there. That lesson came at me again and again.

I shuffled, waiting for Peter to play the owl's song through his iPod. A canopy of stars lit the sky, yet I did not feel the warmth of their light as the darkness around me squeezed in tight, leaving me light-headed.

After a long pause, Peter said, "It's not working." He slipped his iPod into his armpit to warm it. "I think it's frozen."

I gave a half laugh, imagining the iPod on strike, protesting working in such cold.

"Whistle," I suggested, though that seemed a silly thing to do, when the world, so silent, seemed empty. Still, though a whistle wouldn't project as far into the forest, we had to try something. And Peter had a talent for whistling in owls. He had yet to whistle in a Saw-whet, but I had heard him have long hooting conversations with Barred and Great Horned Owls.

Evenings at Peter's house, he often pulled out two plastic chairs and we sat, surrounded by tall white pines. We didn't sip a beer and discuss our day or the future. He whistled the Saw-whet's regular song. *Toot, toot, toot, toot,* he began low in the register, the notes rising slowly. Supposedly this song is what a saw being sharpened on a whetstone sounds like. Since this is a sound I'll never know, I have to trust this is accurate while at the same time wishing the bird had a more poetic name.

At times, lips tired, Peter would pause. I once took over, tooting out the Saw-whet's song. Peter shook his head. I stopped my whistling, embarrassed I couldn't stay on key. The owls would know the difference and never respond. We had yet to have an owl appear, but we didn't allow that they might not be there as we sat and whistled and hoped.

Peter's whistle now pierced the predawn air. At first strong, each toot became fainter as the cold took over Peter's lips. That an owl lurked near enough to hear Peter's half-frozen whistle and would grace us with a visit seemed preposterous. Peter paused for a moment, joined me in staring into the darkness and up, toward a display of light so sharp and clear I felt lucky to be alive. In that moment, I was willing to believe in anything: Saw-whets, God, or the holiness of CBC. In that moment, I became a convert to this annual tradition.

A shooting star traced the arc of the sky, then plummeted as if into the trees. *Wish.* We both laughed out loud. Do stars grant small wishes: may that owl appear. Or bigger wishes: may we be happy together. I made both, overburdening that one blaze of light.

"Wait," Peter said, pointing over the water. "Hear that?"

A weak click, like snapping fingers, emerged from the woods.

I nodded, though I was not familiar with the contact call of the little owl.

Peter gave me a thumb's up. *Saw-whet.* We both jumped with excitement. Then the owl returned Peter's song, its toots evenly cadenced, sweetly resonant. I strained to make out the silhouette of the little bird, a small globe of feathers and killer cuteness, perched on the limb of a tree. I wished those stars shined down even brighter to light the flame of the owl's eye for me.

From the other side of the swamp a second owl piped up. Two! I shifted my gaze back and forth as the owls traded hoots, perhaps discussing these odd two-legged visitors: friend or foe, they might be wondering. The chorus continued back and forth, like hollow chimes in the clear air, while we listened, dumbstruck.

They really were out there, oh ye of little faith.

Two weeks earlier, Peter and I had gathered with a half dozen other bird enthusiasts to watch a researcher, Glenn Proudfoot, band Saw-whets on the Mohonk Preserve. To get the birds to fly into the mist nets, Proudfoot broadcasted the owl's call into the night sky. The loud hoots reverberated through the woods, drawing in curious Saw-whets. It was quite a sight, with over a dozen birds thoroughly tangled in the black mesh, often suspended upside down, indignant yellow eyes blazing. Proudfoot worked quickly, hands bare, untangling the birds as they nipped, objecting to all of this. He then stuffed them headfirst into empty six-ounce tomato paste cans. That's how tiny they are. He had six cans duct-taped together into a triangle so that by the time he had collected his owls, he had a six pack, their feet sticking out of the cans. I might have found this funny except I was so anxious for the owls' well-being I could hardly watch.

Inside a cramped, warmed hut, Proudfoot banded, weighed, and measured each owl as part of a larger project studying Saw-whets called Project Owlnet. While Proudfoot, who looks like the man of the woods with his long dark gray beard and solid hands, systematically went about his measurements, I admired the stocky bird, appreciating its tenacity as it clicked its bill in protest over being fondled, prodded, and flipped upside down. Watching Proudfoot, I was reminded why I never could have been a scientist. Though I understood the importance of Proudfoot's work, and saw how his research might help these birds, that night I couldn't wait for them to be set free.

In a coauthored 2011 article in the *Wilson Journal of Ornithology*, Proudfoot begins: "The Northern Saw-whet Owl (*Aegolius acadicus*) is a common but poorly-understood member of the North American forest fauna." If you are a person, being poorly understood is a problem; for an owl, this is their goal, to keep flying their mysterious path across the globe.

And yet Saw-whets have not always been elusive. When Theodore Roosevelt made his bird list of Washington, D.C., a pair of Saw-whets spent "several weeks by the south portico of the White House in June, 1905." Roosevelt's observation points out the fact that a century ago, Saw-whets were more common, even "household birds," whereas now their population is declining.

It was often thought that owls did not migrate, sticking it out on their territories through the winters. That was the case of the Saw-whet until 1906 when on October 10 an unseasonably cold wind and heavy snow knocked thousands of birds from the sky into Lake Huron, where they drowned. Many washed up on shore, where a certain W. E. Saunders was walking the beach. He recognized the dead birds thick at his feet: Dark-eyed Juncos and Winter Wrens, Song Sparrows and Ruby-crowned Kinglets, and, much to his surprise, twenty-four Saw-whet Owls. This sad discovery set researchers to finding out how far these owls migrate. Proudfoot's study focused on migration, especially route fidelity. *Fidelity*.

I imagined the bird marrying its path north and south, promising to return each season. What love nests in this trail that maps through our skies? Maybe that love is one we should curate, tending to these aerial trails so that the birds can find safe passage.

Too soon, Peter said, "We have to move on."

"Why?" I wanted to stand forever in the dark talking with the owls.

"We have to get Screech, Horned, and Barred. Come on."

We stuffed ourselves back into the car, and I picked up the clipboard, proudly checking off Saw-whet and adding in the number column, 2. As the car started to move I started to shiver, this time from the cold that had taken hold while I was paying attention to the owls and not to my warmth.

I picked up the laminated map of our count circle—which we referred to as the holy scroll—and clicked on an interior light to try and follow where we were going. The count circle we were part of was divided into ten sectors of various sizes, and each assigned a letter. Peter's sector was the long letter H that curved along the western edge of the circle. Unlike other sectors, which held water or preserved woods, this sector offered nothing of wonder, like running water or expansive tracts of uninhabited land. Until Peter came along, it had been an uncounted sector.

Sector H includes winding roads that bisect farms and, at a higher elevation, cut through forests of both hardwood and evergreen trees. Stone houses along with a few trailers punctuate the landscape. Most of the houses appeared shuttered, dark, and at others plumes of smoke rose from chimneys. No one would ever come to this area of Ulster County to bird, and yet every year Peter found special birds working the fields or sunbathing on the limb of an oak.

Peter and I noted our nighttime efforts, the fifteen miles traveled (which felt like thirty), and the owls heard besides the Saw-whets: 1 Barred. 2 Screech. Then, at 7:00 a.m., our daylight charge began. We parked at the edge of farmland, stepped out of the car, and immediately, in the gray light of dawn, the chatter of snowflakes overhead told us we had Snow Buntings. Our first bird of daylight hours.

Just a few weeks earlier I had met my first Snow Buntings in East Kingston off the Hudson River, so these small visitors from the north felt like friends. They yo-yoed through the air, their up-and-down flight forming a wave of white bodies shimmering in the morning light. Circling, circling, they seemed to be searching for the perfect spot to land for breakfast. The white bellies of their compact bodies caught small slivers of the rising sun. Then, like a drape fluttering at an open window, they put to earth, vanishing into a wide, empty field that led toward a house a hundred yards away.

"How many?"

"Twenty?" I hesitated, unused to having to count birds. "Let's go see." I started down a driveway.

"That is exactly what we are not supposed to do," Peter cautioned.

CBC guidelines ask that counters respect private property as it seems birders have a knack for irritating landowners. Maybe people, seeing our binoculars and scopes and cameras, felt spied on.

But at this early hour, I doubted we would be bothering anyone. I took a few steps down a long, dirt driveway. As if I'd pushed a button, a garage door opened and a car, lights on, drove toward us.

"Shit," Peter said.

I admitted this was unfortunate.

The car slowed as the driver lowered his window.

"Do you mind if we walk down your driveway?" I asked. "We're looking for birds. Audubon Christmas Bird Count." I felt that made us sound official.

"Sure, I'm a birder too," the man said. Without further conversation he moved on. I tried not to look too pleased as I continued into the field to count twenty Snow Buntings picking through the wide field at dawn.

For the morning hours, we drove the twisting country roads, windows down, heat pumping into the car to offset the cold. From time to time we pulled onto the shoulder of the road or if there was no shoulder just stopped, so little traffic to disturb us. Windows down, we counted the expected species of Chickadees and Cardinals, Blue Jays, and Pigeons or Rock Doves, winging out of barns. Some birds we counted as we drove, the Red-tailed Hawks perched on a limb overlooking a field or the Canada Geese grumbling in fields. I dutifully tallied the numbers, often overwhelmed as Peter called out, "add three more Crows" or "Waxies, five."

"Where?"

"That bush we just passed."

How could he hear or see Waxies while driving?

We followed a narrow road that led to a dead end, a swamp framing the road. Despite my time studying the map, and as if to refute my sense that I am directionally adept, I had little idea where we were.

Peter parked the car. "Watch this." We both got out, faced the swamp as if it might be a stage.

He *pished* into phragmites and cattails, which canted at odd angles. A few snags, bare of bark or leaves or life, framed the dried grasses. Frozen water circled the bare trees like ice skirts. The place seemed devoid of kinetic life, holding its breath until spring and warmth. I knew that by midspring this little swamp would be alive with flies and moths, peepers and caterpillars, and all of the birds that chase bugs for a living. For now, silence.

Peter *pished* again, his harsh alarm call waking up the cattails.

A sparrow sat up, clutching a hollow reed. Gray cheeks with a black eyeline. Gray clear chest. Rufous body. Just as fast, it dove back into its frozen existence.

"Swampy," Peter glowed.

I was happy to see the Swamp Sparrow, a bird whose staccato song I now associated with the hot summer months in the Tivoli Bay, and I shared Peter's happiness in seeing the bird. Then I soon started to worry if the bird was OK and did it have enough to eat. I wondered aloud why it had not headed south with its comrades, to a place where bugs were plentiful, where life, so much warmer, would be easier.

"Dunno," Peter said. "It seems fine."

At first I thought Peter a bit callous, but I now understood that he had faith that nature knew its course, that birds followed food and took care of themselves. Peter trusted that nature did what it needed to do. My worry seemed a bit presumptuous: Why would I know better than the sparrow where it was good to live?

We slammed car doors and raced off, as if we had somewhere important to be, the thrill of that lone, maybe cold, bird pushing us past houses surrounded by pastures where cows clustered together against a fence, or at one farm, alpacas. I wondered, as I often do, what people did with their lives here in the woods. Maybe they worked in Kingston, a half hour drive away. Or they had figured out a country life, growing their vegetables and canning tomatoes to make it through the winter.

"It's a balmy twenty-two," Peter said as I turned up the heat. He was dressed in a black snowmobile suit with thick insulation; the nylon swished when he walked. "You aren't wearing enough clothes."

I chafed at Peter's suggestion that I didn't know how to dress for the cold. After all, I was the one who had survived storms and negative fifty-degree temperatures when I travelled to Antarctica. I had on as many layers as was possible that would still allow me to get in and out of the car.

"I need to move," I complained. "Get the blood circulating."

"Sorry," he said.

I had been right to worry that this was car birding, but Peter was also right that a holy enthusiasm suffused the day, which made me soon forget my ice cube toes.

Soon Peter did let me walk. We got out to comb a wide corn field that a farmer had given us permission to search.

"What are we looking for?" I asked.

"A Lapland Longspur."

What did a Longspur look like, I wondered.

"*Anything* is possible. It's Christmas Bird Count." Peter's sense of the wonder of the day was a tonic. For him, it was a free-for-all day, one where special birds lurked in every bush. I didn't doubt him. And at the same time, I knew the birds didn't care about our holiday. They had been living their lives, not waiting for us to come and say hello and mark down their existence on the tally sheet.

"We should have Horned Larks, or Pipits." That seemed a more reasonable expectation in this field. We moved along the rows, guided by short dry cornstalks, kicking up Song Sparrows and White-throated Sparrows.

Peter started to sing, perfectly in key:

When a good bird comes along
Is it a pipit?
When it sings its little song
Does it say "pipit"?
Get it straight
Move forward
Move ahead
Try to detect it
It's not too late
To find a pipit
Pipit's good.

Devo. Punk rock. The 1970s when Gerald Ford, who we mocked for not being able to walk and chew gum at the same time, was president and Peter and I were both in high school. Would we have liked each other in high school? During those awful adolescent years, I made it my duty to push boys away. I doubt I would have noticed Peter unless he rock climbed, and I imagine Peter attracted to girls not like me with my long unkempt hair and dressed like a tomboy in white painter's pants and a broad-striped rugby shirt—the climbing outfit of the time.

Despite Peter's song, we saw no Pipits. Or Longspurs.

Back in the car, we stopped near bird feeders, counting Juncos, House Sparrows, and Tufted Titmice. Then near ten in the morning, our day already so full, Peter stopped at a convenience store, in need of coffee and chocolate (the two things I managed to not bring). At the edge of the parking lot, I pointed up at a small raptor.

"Merlin," Peter said as if this might be the most expected bird to find perched in a tree hanging over a convenience store.

Grinning, I watched the falcon while Peter took photos. The bird didn't budge when Peter finally left to get his coffee. I continued to admire its mottled chest, long subtly striped grey and black tail. The bird returned my gaze with its piercing eyes. Like all falcons, it appeared pleased with itself as it sat, content and dismissive, while people came empty handed and left with donuts and coffee.

To have such a neat bird hanging around that unexceptional parking lot, to have good birds throughout our supposedly uninteresting sector told me that there are good birds everywhere. I wondered what it would be like to live every day on high alert, as if it were CBC. What might I see?

Midday Peter said, "We need a Red-breasted Nuthatch." I held the checklist of birds on a clipboard; Peter held the list of birds in his head. He turned the car around, driving out to deep woods, to a stand of pine trees where he knew a Red-breasted Nuthatch should be. He drove along at ten miles an hour, windows down. From the edge of the woods, the bird let out its abrupt yank of a song.

"Amazing," I said, shaking my head.

The rest of the day unfolded in a daze of counting, walking, stopping, and driving, always with the windows down so Peter might hear some faint, important chip note or song. The wind on my face made my cheeks red, and my nose felt cold all day.

We did not talk, except to discuss birds. A week earlier I had moved out of Peter's house. It was a commonplace domestic mishap. "That's not going to work," Peter said as I lay kindling in the wood stove. I had lived for years using a wood stove as my sole source of heat, so I'm handy at starting a fire. I just wasn't doing it as he would. This I know: those of us who love fires are particular about how they are built and tended. His irritation, though, ran beyond building a fire.

"You don't want me here, do you?" I should have noticed the signs this wasn't working. But like an opossum, naïve and slow to learn that the road is a dangerous place, I ignored the wariness in his eyes as we stood side by side brushing our teeth.

What he said in response skirted the idea that it wasn't my fault; even though sharing his house had been his idea, he wasn't ready for a domestic life. I understood that he was too fresh out of his marriage and that it made no sense to jump back into something that looked, spookily, marriage-like. Neither of us yelled or said a mean word; I've never been one to fight. I packed my bags and was gone by midmorning.

The only thing I left behind was a pumpkin pie I had baked the day before, the smell of sweet and cinnamon filling his house. It's odd how the small things take on such meaning; I realized as I sat in my kitchen in Tivoli wishing for a slice of that pie that in my own small way, I *had* been trying to make a home with Peter.

My little wooden house smelled damp, unlived in. My fault or not, I was hurt and upset that a relationship that I had felt to be so promising had ended so fast.

"Let's be adults about this," Peter said, when I suggested we talk about how to end our relationship. I wanted to sort out our feelings for each other from our love of the birds. I wanted us to stay birding partners.

I liked how "being adults" sounded because often the way we behaved—running through fields chasing birds—made me feel like we were kids. My passion for birds made me self-conscious, not because of my dorky appearance with binoculars strapped to my chest and practical pants bagging in the breeze, but because my feelings were obviously overwhelming. I acted like a teenager in the throes of a first love, forgetting sleep and food, as I chased birds.

I agreed to act like an adult, which for Peter meant giving the relationship another try. Perhaps with me back at my house, we could wind the clock back a few months, to when our relationship held such hope. So we moved forward, a little doubtful, a little half-hearted. What kept us focused outward, forward, onward were, of course, the birds. During Christmas Bird Count the pursuit of birds created an odd cocktail of exhaustion and elation, the same irresistible tension of new love. I easily translated these feelings to Peter; after all, here was a man who could whistle in Saw-whet Owls in the dark.

At 4:30 p.m., we packed up and drove to Kingston, where all of the other counters in our circle gathered. There were thirty people in the field that day, ten teams total, all adding to sixty years of bird data for this circle. All of us were tired and a bit high from the long hours, the constant focus. Some of these birders I had encountered on walks, while others I knew from reputation. There was Mark, whom I had met on Slide Mountain and on several other walks, jubilant from his day birding the streets of Kingston; his wife Kyla had organized all the food. And I was finally meeting Frank, who wrote lovely emails from his home in Saugerties, where he searched the Hudson River for birds. He signed his posts Denny Droica (say it fast: he's *Dendroica*, at the time a genus of colorful wood warbler). With him was his girlfriend Deb. There were Wendy and Jess, whom Peter referred to as the Tocci twins, even though they are not related. All of these people were, in my mind, stars, birding heroes. To meet them there, a bit disheveled, a bit tired, made me smile with relief: they were just human.

Peter referred to the stone-walled room owned by the city of Kingston where we gathered as the dungeon. And he wasn't wrong. Boards shuttered the windows against winter storms, giving the sense we were kids who had broken into an abandoned building. Dusty, exposed light bulbs weakly lit the long table. No one seemed to care that the room was barely warmer than outside as we milled about, eager to share, bragging a bit, joking about the weather. Word spread of a few unusual sightings, like a Killdeer, a Broad-winged Hawk, a Ruby-crowned Kinglet, Wood Ducks, and Ring-necked Ducks. The bird banter reminded me of evenings after climbing, standing around the local bar sharing what we had done that day. The point was to brag while remaining casual, as if I'd just waltzed up the Yellow Wall.

"Don't say anything about the Saw-whets," Peter whispered to me with a wink, and I smiled.

We heaped paper plates with hot food—baked ham, mac and cheese, turkey and mashed potatoes as if at a Thanksgiving feast. After our day of nonstop eating I was not particularly hungry, but I dug in, sitting elbow to elbow at the long table, one big family of birders drinking beer and cider. I imagined we might sing a song of thanks to the birds; instead, we ate. Then came the compilation.

Leading us was Steve, who had been the CBC compiler for this count since 2003. I knew Steve mostly through his precise, elegant emails detailing sightings in Ulster County or offering an account of his day spent counting birds for the Winter Survey at the Esopus Bend Nature Preserve. In person, Steve is a quiet man, with long hair and smooth skin that makes him ageless. On the few walks that I had shared with him, I noticed that he misses not a click or peep or glimpse of a wing.

And then he started the count, all of us falling silent. Steve called out each species, beginning with Canada Geese. In turn, each sector leader offered numbers seen.

"Wood Duck."

"None."

"None."

"None."

"Three."

"Five."

"Black Duck."

"None."

"None."

"Ten."

"Three."

"Thirteen."

The call-and-response had a cadence at once joyful and purposeful. At times someone would doze off or was not paying attention. "Sector F? Ring-billed Gulls?"

With certain species, people would comment. "Five eagles? Remember when we got none." Or when no one reported a Ruffed Grouse Steve pointed out that in the 1950s, every sector counted at least one, usually several. "They're done," Frank said ominously. With the Red-bellied Woodpecker—a total of sixty-three, and seen in every sector—Mark remembered few or no birds in the 1970s. Some of these fluctuating numbers can be attributed to the flow of life, the flow of death. Some of it is surely due to changes in our environment. Either way, I envied that these people had a deep knowledge of this area, knew the populations of birds as if they might be their cousins. It made me realize that learning birds doesn't happen overnight, can't be taken in during a semester, isn't a discrete skill that you perfect, but a lifelong process. Everyone in the room was still learning their birds.

I waited, eager, as Steve moved through the owls. Screech. Barred. Finally: "Saw-whet?" Each sector echoed none.

"A pair," Peter said as calmly as he could.

I sat up, ready to stop the compilation to tell the story of Peter whistling into the dark and yet, shy in this crowd of birders, I held back. Mumbles of "good, good bird," rose around the table along with questions of *where* from others.

"That might be considered a rarity," Steve said, before continuing with the list. Peter frowned, as that would involve filling out some paperwork.

When Peter reported no Great Horned Owls, Steve turned. "How come you didn't find any Great Horned?"

Peter dropped his head, a gesture of mock shame, "I guess I'm not very good at this."

Steve hesitated for a moment. He turned back to his computer screen, ready to move on. A smile appeared and then, still focused on the screen, that smile spread, his face lit up as he started to laugh, his hunched shoulders moving up and down. Soon, we were all laughing, from the wonderful absurdity of Peter's comment, from the pleasure of birding all day, of finding a Saw-whet and not finding a Great Horned, of sharing this deep, odd love of birds.

DON'T MOVE

—

Paris, France, and Pennypack Preserve, Pennsylvania

When John Burroughs invited readers to know the birds, he encouraged them to get a guidebook, John James Audubon being the best. "His chapter on the wild goose is as good as a poem. One readily overlooks his style, which is often verbose and affected, in consideration of enthusiasm so genuine and purpose so single." Since Burroughs was pretty verbose himself, his critique of Audubon's style must be taken seriously. I'm not sure there's much goose poetry in the account of the Canada Goose in the *Ornithological Biography*, though Audubon does offer a fantastic description of the birds mating that involves darting fiery eyes and quills that shake with the intensity of emotion. But most people don't turn to Audubon for his prose; they admire his paintings.

To create his drawings, Audubon did not just shoot the bird, string it up in a somewhat lifelike manner, and then reproduce it on the page. He poured into his art hours of observation of the bird's feeding and behavior in the wild. He wanted to know the bird alive and to capture it on the page. In my judgment he often falls short; his birds look like just what they are: stuffed dead birds in often odd poses. Though Burroughs admired Audubon and his art, he faults the spirit of some of his drawings: "His bird pictures reflect his own temperament, not to say his nationality; the birds are very demonstrative, even theatrical and melodramatic at times. In some cases this is all right, in others it is all wrong. Birds differ in this respect as much as people do—some are very quiet and sedate, others pose and gesticulate like a Frenchman." Yes, sometimes Audubon makes his birds too *French*.

That Jean Jacques was French is a fact that Burroughs never lets the reader forget. Audubon's father was a sea captain, and his mother was from Saint Domingue (now Haiti); she died shortly after his birth. Jean Audubon (père) took his son home to Nantes in the western part of France where his wife not only accepted this illegitimate son, she spoiled him. Audubon claimed he had carte blanche at all of the confectionary shops in town. His stepmother didn't insist that he learn math or even that he attend school. Free to roam, the young bird enthusiast spent his days tromping through fields where he created his first drawings of birds. "He speaks of his early intimacy with Nature as a feeling which bordered on frenzy," wrote Burroughs. In 1803, only eighteen years old, Audubon arrived in America filled with an "intense and indescribable pleasure." He wasn't coming only in search of birds; he was, above all, avoiding conscription in Napoleon's army.

Though Audubon is now celebrated as the bird artist, his early path was rocky in many ways. After moving to his father's estate in Mill Grove, Pennsylvania, Audubon soon married Lucy Bakewell; she gave birth to two sons and a daughter, who died in infancy. Audubon had difficulty supporting his family as he had no real knack for business. He was swindled in various deals, and in his dreamy way he misplaced money. When his work failed in Pennsylvania, he moved west, to Kentucky.

One of Audubon's business partners in Kentucky was George Keats, who had journeyed from London, England, with his wife. The Audubon and Keats families lived together in Henderson, Kentucky, before moving to Louisville, and the two men embarked on several business ventures, none of which was a success; most left Keats destitute. Keats's grandson wrote that Audubon sold his grandfather a "boat loaded with merchandise, which at the time of the sale Audubon knew to be at the bottom of the Mississippi River." This statement isn't true in fact but more in spirit; Audubon was reckless not only with his own money but with that of others.

Keats's brother was none other than the romantic poet John Keats, who had no love for his brother's business partner or for his wife: "Give my compliments to Mrs. Audubon and tell her I cannot think her either good-looking or honest," he wrote to his brother's wife. "Tell Mr. Audubon he's a fool." Keats's letter is a lyric of another sort, and a great reminder that one person's hero is another's villain.

In 1819, Audubon declared bankruptcy. Rather than wallow in despair, he recognized: "I *had* talents, and to them I instantly resorted." Drawing portraits for five dollars apiece, he began to right himself financially. His work became so popular that he claimed a grieving parent had him disinter his son to draw his portrait. Age thirty-five, Audubon had his moment of conversion. From then on he dedicated himself to the birds, drawing their portraits.

Knowing Audubon's wayward ways, I wondered if perhaps I could blame some of my dreaminess on my French blood. My mother was French, my father American. I hold two passports, speak two languages, and love both countries. I visit France and my sister, who lives there, as often as possible and usually at Christmas. It took a while to convince Peter that Christmas in Paris was a good idea.

If Peter was an optimist in relation to birds—the Saw-whet is out there; there must be a Lapland Longspur in with the larks; the Ivory-billed Woodpecker still flies—I had learned he was a pessimist in life. There was never enough time or money or maybe there wasn't enough love to do all of the things I wanted to do. Still, with the promise of new birds, Peter signed on. We arrived in Paris fresh off our Christmas Bird Count and three days before the other Christmas.

Peter had never visited Paris, so I was eager to show him the churches, squares, and museums I most adore: St.-Germain-des-Prés, the Places des Vosges, and the Musée d'Orsay. Above all, I wanted to show him my Paris, the American Church where I was baptized in 1970 (a sort of afterthought baptism as I was nine years old), the all-girls school I attended in third grade on the rue Sarette, and the Parc Montsouris where Becky and I played as kids after school, hopping through our Chinese jump ropes or knocking our *cerceau* in circles around the lake. We often brought leftover bread to feed the ducks and the elegant, mean, Black Swans. Signs now forbid this casual feeding, but it's clear that most of the birds still expect to see pieces of a stale baguette land in the water for an easy meal.

I wanted Peter to love this city; Paris was not making itself lovable. Gray skies greeted us every morning and thirty-four-degree rain came down at such regular intervals we never dried out or warmed up. One website, Birding Paris, admitted that no one comes to Paris to bird—really, what an absurd idea! Yet, since Peter cared less for sights other than the birds, that's what we did, slogging out every morning on a birdy mission.

This meant that we made a tour of the parks of Paris including the Tuileries, the Luxembourg, the Bois de Vincennes, and the Bois de Boulogne. If you know your Paris parks, you know there's a big difference between the manicured Luxembourg and the ranging Bois de Boulogne near the periphery of the city where prostitutes work the narrow roads and people set up their camping caravans in the summertime. At each of these parks we found common birds that for us were marvels. In the Tuileries we met the Black-headed Gull letting out a yelping call and the Great Tit, a bird the size of a chickadee, which peered down at us revealing its black cap. That both birds had the blocky Louvre as a backdrop to me made the birding more fun.

Did Peter note the Louvre? Maybe. But did it matter? Peter was rewriting what mattered. The gulls mattered. I soon let go of my Paris agenda and gave over to the birds of France. Rather than showing Peter my Paris, he was showing me the bird life of this city I thought I knew well. Still, I had to marvel that Peter was perhaps the only person I knew not to be seduced by the idle pace of Paris life, and that we became the only couple in the history of the world to not loiter on the quais of the Seine for a kiss.

We were in more ways than one out of step with Paris. Since Peter had not anticipated such dismal weather (how could he?), he had not brought enough warm clothes. So he wore my brother-in-law, Olivier's, bright red downhill ski suit. In addition to this snowman look, Peter carried his camera wherever we went, relying on his photos to help piece together identifications of these foreign birds. Out in a swamp or tromping across a field, he naturally slung twelve pounds of glass at his side. In a city, though, a 500 mm fixed lens wrapped in a camouflage sleeve stands out. "Paparazzi?" someone called to us the first day. Often people stared, trying to figure out if there might be a political figure or movie star nearby.

"I was ridiculously fond of dress," writes Audubon in his essay "Myself." "To have seen me going shooting in black satin smallclothes, or breeches, with silk stockings and the finest ruffled shirt Philadelphia could afford, was, as I now realize, an absurd spectacle, but it was one of my many foibles, and I shall not conceal it." Audubon would not have approved of Peter's snowman outfit, and neither did some Parisians who, in their inimitable manner, stared with disdain.

On the day before Christmas, we navigated the muddy, snowy paths in the Bois de Vincennes, which brought us a Mistle Thrush and a Great Spotted Woodpecker, a bird that does have two great white spots on either shoulder, a red band at the back of the head, and red underpants. We spent hours in the park, combing the woods and scanning the lake where we found a range of new birds: a Ruddy Shelduck, a solid duck with a burnt orange body, black necklace, and cream-colored head; a Barnacle Goose, similar in appearance to a Brant with a white face; and a Grey Heron at the shoreline, a match to our Great Blue. Only in making comparisons could I begin to orient myself within the French birds.

If Peter felt out of his depths navigating the metro or finding meaning in the conversation of French-speaking people, I was out of my depths with the birds. If Wyoming felt disorienting, a whole new continent of bird species was overwhelming. I left Peter to piece together the identifications of the birds we found. This was the first time I had seen him carry a guidebook in the field, the first time he hesitated with an ID. It was good to see him go through the motions I went through daily at home, to realize that in fact his knowledge of birds wasn't heaven sent but rather a process that involved studying the guidebook, then trial and error in the field.

When my frozen, wet feet couldn't take it anymore, we moved toward the gates in the Bois de Vincennes, making a detour around a homeless encampment. Tents in blue and yellow wedged between the rows of plane trees. A bicycle propped against a nearby rock, while a few small fires sputtered, letting off the homey smell of burning wood. A child poked her head out of an unzipped door. It could have been a campground except that this was a pretty rough camping experience. If I had become cold in a few hours out of doors, I couldn't imagine a life in a tent there in the slush of winter.

At the entrance to every church in Paris men and women sit, draped in layers of clothing while holding out a cup or hat, daring tourists or the faithful to be truly Christian with a euro or two. I expected these homeless people when we frequented my Paris. A dozen tents, though, took the misery to another level. And those rain-soaked tents set the frivolousness of our birding into uneasy relief. This was not the first time I had meditated on the privilege of birding, but it was the first time I felt confronted by it. As I ticked through unsettling emotions, my solace was that had I not been birding I would never have seen these homeless people. Birding showed me not just the birds, but the world that we share with them. Witnessing the lives of these people in their tents was important, a step in understanding; only with understanding can change begin.

I often ended a day birding at once elated by all I had seen and also troubled that once abundant birds are now teetering toward extinction. The ways that we have challenged birds left me angry and frustrated. Great joy comes, as we know, with great pain. Birding in a city was leaving me less elated and now mulling over the ways we have challenged our own species. Not that I hadn't thought about, read about, talked about these challenges before, but the birding lens added something.

When looking at a bird, nothing else matters; there's no past, no future, only the perfect present. It's much like being in the presence of a great love, nothing but being there, together, matters. To train that eye on people, I maintained that same focus, but the present wasn't perfect. Neither were the future or the past. What remained was a sense of affection as, for a moment, I saw these people who were as unknown to me as the birds. The birds led me to feeling a greater compassion.

Audubon traveled widely to gather his bird specimens. In the early nineteenth century this was not a snap proposition. His family carried on without him for long stretches of time, once, according to Burroughs, for five years. "One often marvels at Audubon's apparent indifference to his wife and his home, for from the first he was given to wandering," writes Burroughs, who critically noted that Audubon's wife had to work to support the family. Stated another way: Audubon was selfish; the search for birds is selfish.

The source of selfishness is varied: ego, carelessness, or cruelty. But some selfishness is also a sign of protection, cocooning into the self or a world for any number of reasons: to heal, to gain strength, to survive. Rock climbing, now birding—I acknowledge that both are selfish activities, and neither does the world much good. And yet I knew that climbing had made the world bearable for the young me. I saw that birding made me more fit for the world, gave me resources I hadn't had before: greater patience, and a generosity born from my concern for the fate of all birds.

Still, on this trip, a self-consciousness about our selfish birding took over, triggered by seeing the homeless in the Bois de Vincennes. In a practical way, our birding time meant that I wasn't helping Becky much with the holiday preparations. I wasn't spending time with my nephew and niece, whom I normally doted on.

"We've seen her every day," Peter corrected.

He was right, we did end each day chilled and tired in my sister's apartment in the 14th arrondissement, overlooking the Parc Montsouris. By then the shopping had been done, the dinner half prepared, and the kids were immersed in games with friends.

As I joined Becky in the kitchen on Christmas Day to cook the evening meal we would share with Olivier's parents, I joked, "You can take a birder to Paris, but you can't lead him to a museum."

Becky smiled. My sister had only heard of, not been witness to, my birding passion, which was only eight months old. I could tell it left her a bit puzzled. Every morning she watched, incredulous, as we bundled up to venture out in the dismal cold.

"Do you actually enjoy looking for birds in this weather?"

I hesitated.

Peter, who had been napping in the living room, suddenly appeared at the kitchen door. "I need to hear the answer to this question," he said with a bit of a smile.

"I do," I said, aware that was not entirely true. If Peter hadn't been there, I would have been laughing through a Charlie Chaplin rerun in one of the small art theaters near St. Germain, or studying the paintings of the revolutionary-era painter Jacques-Louis David, whom Audubon claimed to have studied with (this was pure résumé padding).

Behind Becky's question lurked what really puzzled her: Did I like this guy who was so obsessed? I didn't see Peter's birding drive as a negative;

I only saw the commitment and devotion; it was something I admired. I have always fallen for people passionate about what they do, whether rock climbing, writing, or saving our rivers. Peter's focus was part of his appeal. In Paris, however, I recognized that passion had mutated into obsession. That was a lot less sexy.

I'd been driven by just such an obsession when I climbed. The often inconsiderate me, myopically turned toward the cliffs and ruthlessly cutting out the world, was not someone I remembered with much love. And yet I accepted Peter's ruthlessness because I saw how it was necessary if he was going to give his life to the birds. In other words, I was fine with the fact he was sometimes an asshole.

Over dinner, Olivier's mother, elegant and in her seventies, commented that Peter looked like Van Gogh. "Tu ne trouves pas?" Fanfette nudged me, giddy. I had never made the physical connection, yet I saw that she wasn't wrong. It wasn't just the trim red beard, though Peter's mustache was slim and discreet in comparison to the artist of Arles. They had the same high forehead, slender nose, eyes that didn't waver. Peter winked at Fanfette. *I've fallen for a madman*, I thought.

It was to Paris and to London that Audubon turned in 1828, once he had gathered his specimens and drawn them to create his *Birds of America*. While in Paris, Audubon met many great men, among them the father of paleontology, Georges Cuvier. I can't resist offering Audubon's description of this important French naturalist:

> Age about sixty-five; size corpulent, five feet five English measure; head large, face wrinkled and brownish; eyes grey, brilliant and sparkling; nose aquiline large and red; mouth large with good lips; teeth few blunted by age, excepting on the lower jaw, *measuring nearly three-quarters of an inch square*. (italics added)

A bird ID transposed into a sketch of what sounds like a rather grotesque human. Audubon could see his environment, human and natural, clearly, yet at the same time lived in a world of his own creation. There are five birds that Audubon drew that leave ornithologists puzzled: the Carbonated Swamp Warbler, Townsend's Finch, Cuvier's Kinglet, the Small-headed Flycatcher, and the Blue Mountain Warbler. What *was* he looking at? And did Cuvier really have astonishingly large teeth?

If Audubon had enjoyed his glamorous time in Paris with the Baron and others, Peter did not enjoy his middle-class holiday. To try and improve the situation, the day after Christmas I suggested we visit the forests of Fontainbleau.

When I lived in Paris in 1984, I had often driven the hour south to climb the boulders that dot the landscape. I had loved my weekends at Fontainbleau, with climbing friends but also with friends who had never climbed before who brought their kids along. There was something for everyone, the boulders playful or challenging. But it wasn't just the climbing that I enjoyed, it was how civilized it all was, how climbers brought little carpets to wipe the sand off their feet before launching into a climb. And then how, sometime around noon, everyone stopped to share lunches made up of baguettes and cheese, saucisson or pâté, and sometimes even bottles of wine. Despite my love of food, I had from my early days associated having a real adventure with hunger, and those days in Fontainbleau rewrote that story: you could eat well and have a good time.

Peter was more than game for this idea—anything to get out of Paris. Plus, somewhere online he had found a site that promised Black Woodpeckers in Fontainbleau.

"Did they say where in Fontainbleau we might find Black Woodpeckers?" I asked, knowing that the forest spreads to almost 250 square kilometers, encompassing several towns. For once I was the pessimist. "Seems unlikely we'll find one," I said, as late morning we sped south out of the city in Becky's car.

We scooted out to Barbizon, a cluster of stone houses, where the painter Jean-Francois Millet had created his studio. The sun was not fully with us, but at least the rain had stopped, and some of the gray we had experienced in Paris had lifted. From Barbizon, we took to the trails that spread through the Gorges d'Apremont. The muted woods calmed our city nerves as we breathed in crisper air. The soft sandy soil beneath our feet made it feel like we walked in slippers as we followed the blue blazes, meandering through dense woods and around startling rock outcroppings that rose ten or twenty feet. I looked at the boulders with affection, imagined pulling on a sharp hold or stepping high to get off the ground before moving vertically.

We didn't give our direction much thought, the trail well marked. We walked side by side, stopping when we heard a chip or pip that might lead us to a new bird. Though the trail was quiet, I was happy to just move, to stretch our legs, to bird in a way that felt familiar. After two hours of drifting along the flat trail, we sensed we were not looping back to town as we had anticipated so we decided to retrace our steps.

As often happens birding, I was alert on the way out, then relaxed on the return. I let my thoughts drift, rather than staying focused outward. I floated surely into glum thoughts about our odd, not entirely enjoyable holiday.

Then Peter started to sprint down the trail. "Whoa, whoa, whoa, whoa."

I had never seen him move so fast. He made a sharp left into the woods. "Woodpecker!"

I followed, tracking the movement of a black bird the size of a crow. Amazingly, it landed, glued upright to the trunk of a large tree. *A Black Woodpecker*. It tilted its head, showing its white eye, alert to the world. It had a stunning red cap that contrasted with the all black body.

"Don't move," Peter whispered as he raised his camera. I wasn't sure if he was talking to me or to the bird, but I froze in place, holding my breath as I looked at the Woodpecker and heard Peter's shutter release. I hoped in the dim light that he had captured the bird.

So fast, the Woodpecker flew off. We both stuttered out a laugh taking on the rush of excitement that comes with glimpsing a special bird. We loitered in the woods for a while, knowing the bird was gone yet still expectant. Soon, we straggled back to the trail. A lightness filled me as the woods became darker. One bird, and our sense of our trip transformed into a success.

Late morning on the 31st of December, we were back in the United States, a little jet-lagged after our flight the day before. We sat in Peter's sister Mary's living room in a small town outside of Philadelphia. Mary's house had been the family home, where Peter grew up. Around us swirled his five siblings with their spouses, children, and their children's children. Coffee brewed, someone tussled the skinny dog, a baby made noises from a corner. I was having trouble keeping track of who was who. With my family we'd been swimming in a pond, and now with Peter's family we were swimming in a raucous ocean.

"You had a great trip?" Mary no doubt had images of the Paris most people visit, filled with cafés and cathedrals, with afternoons spent strolling along the Seine.

"We saw a Black Woodpecker." Peter sat in the comfy couch polishing his binoculars.

"Oh, stop."

"We did."

Mary rolled her eyes.

"Let's go find some birds," Peter said.

"Should we stay and help out around the house?" I wondered aloud. Mary had a long list of things to do as that evening was her oldest daughter's wedding. In a lovely twist of circumstance, Mary's daughter was marrying one of my former students. Our worlds overlapping in this way made me feel like I already belonged in this large, gregarious family.

"Mary has it all figured out," Peter said, grabbing his camera.

"Go," Mary encouraged. No doubt she knew she couldn't stop her brother.

We drove ten minutes to the Pennypack Preserve, 852 acres of meadows and woodlands along the Pennypack Creek. There, snow covered the ground of a wide empty field, a white blanket that stretched untouched to the horizon. A Northern Harrier glided across the wide field, gracefully suspended in the silent air.

About one hundred miles from where we walked, Jean Jacques Audubon had roamed when he first arrived in America in 1803. Alexander Wilson also spent time on the Pennypack Creek. This area was rich in history and in bird history.

"We should find a Long-eared Owl in those trees," Peter said.

White pines stood packed in tight rows at the top of a gentle hill. Peter brushed his way into the trees at one row while I slid in fifteen feet down. The smell of pine floated into the air as I took a moment for my eyes to adjust to the speckled lighting. Then I turned to peer up the trunk of the first tree, thinking, with a smile, *here we go again.*

Peter and I spent a lot of time looking for owls, and the ratio of hours spent to owls found is something like a thousand to one. In other words, for all of our time looking into trees, I had yet to find an owl peering down at me. Still, I searched. And I loved that search. Because owling is different from birding. Owling, I had learned from the Saw-whets, is about belief and hope: seeking what is never to be found, yet still taking on the quest (and maybe getting lucky). But it is also about stillness, not moving quickly, or for long moments not moving at all as eyes travel from limb to limb, focusing to make out shape, texture, anything different from the pine needles and bark. If I abandoned my sense of a higher purpose, I laughed to myself that looking up the trees felt a lot like peeking up someone's skirt.

Finding Long-eared Owls has never been an easy proposition. Arthur Cleveland Bent, who from 1919 to 1968 published *Life Histories of Familiar North American Birds* in twenty-one volumes, described finding seven Long-eared Owl nests. This seems at once a big and small number. Big because who finds even one in a lifetime? And small because this is a lifetime of work.

Disturbing an owl on his or her nest is something you don't want to do. "I know of no bird that is bolder or more demonstrative in the defense of its young," wrote Bent, "or one that can threaten the intruder with more grotesque performances or more weird and varied cries." I went looking for other descriptions of those weird calls and found that Bent was not wrong. Edward Forbush offers these descriptions in his *Birds of Massachusetts*: "the preliminary notes of a caterwauling contest . . . like the snarl of an angry feline"; "a peculiar whining cry much like a very young puppy"; and "a soft wu-hunk, wu-hunk slowly repeated several times and a low twittering, whistling note like 'dicky, dicky, dicky, dicky.'"

At another nest, Bent wrote, "Both parents were very demonstrative, flying about close at hand, alighting in the tree close to me, threatening to attack me, and indulging in a long line of owl profanity."

None of this owl noise disturbed the air. In fact, a meditative calm defined our quest. It felt good to be home, back on American soil with American birds and American winter sunshine after our long gray week in Paris. I was relieved to end my worries that we were having a bad time; Peter and I were once again united in our search. Owl thoughts took over. I silently spoke to the owl, sang an owl song, shared what little owl profanity I knew.

After half an hour of quietly walking, looking, hoping, Peter waited on the edge of the woods. "Let's go," he called.

Something in me did not want to give up. "Just a sec," I half whispered. How could Peter stop? At what point do you give up when you are certain there is an owl? *Just one more tree.* Then another. And another.

I knew we couldn't dawdle there in the trees. We had to shop for the wedding as I had managed to forget shoes to wear with my black dress. In less than twelve hours I would be out of long underwear and boots and sporting ridiculous high heels. "Give me your arm so I don't fall over," I said, reaching for Peter, as that night I stepped into the beautiful Bryn Athyn Cathedral, which soared lit into the New Year sky.

I started to turn toward the field when a glint of yellow caught my eye. I stopped, focused. The yellow took shape, two round eyes. The big ears poked out, comic-book style, a rabbit of an owl. Its feathered feet clung to a branch not more than six feet above me. There it sat, blending into the trees, looking a lot like a broken branch, just as natural as can be. We shared a moment of owl to human communication, so direct, so uncomplicated. My moment there with the owl did not make me feel selfish but rather that nothing could be more important, beautiful, exciting than looking into this owl's eyes and sharing some owl profanity. *Don't move,* I willed the bird.

"Hey, Peter," I called in a low, breathless voice. "I have a gift for you."

No Other Everglades

Everglades, Florida

In the heat of a Florida day in March, I plunked down under a palm tree at Flamingo, the end of the road in the Everglades National Park. While I snoozed, recovering from days of only half sleeping in the tent and waking early to bird, Peter wandered off with his camera.

Soon enough—had he really been gone at all?—Peter roused me. "You have to come see this."

I peeled myself from my resting spot, following him to the edge of the parking area where a flock of black birds milled about in the short grass.

"These are Brown-headed Cowbirds," I said. I also knew that probably a special bird had joined the cowbirds. Peter had a knack for finding the least interesting special bird. On this trip, we had made an unfortunate detour into the parking lot of a Burger King in the city of Homestead, which is perched near the entrance to the Everglades. There, mooching off of the dumpster food, was a Common Myna, a blunt brown bird introduced from Asia and one hardly worth the smell of the greasy fries. But a new bird for us both.

For the moment there in Everglades National Park what stepped in front of me were a few dozen Brown-headed Cowbirds with their black bodies and brown mantles. They had a bullying way of moving, chest first. Or maybe I imagined that, given what I knew about how they laid their eggs in the nests of other birds, like the Kirtland's Warbler. Cowbirds are *brood parasites*. The species that accepts a cowbird egg is called the *victim* and the one that raises a cowbird young a *host*. It's hard not to overlay these words with our own moral sense, which becomes mixed here: we want to save the victim, but a host is not a victim, a host is welcoming, offers tea and some cookies.

"I didn't come to the Everglades for a cowbird," I said.

"Dint?"

"Did not," I corrected myself. Making fun of my *dint* was Peter's reminder to me that I grew up in central Pennsylvania, that despite international travel and a lot of education I still had a local edge. What he could not have known is that I liked sounding like I came from the place I grew up, wished in fact that I had more of these verbal ticks, said *iggle* for eagle or *crick* for creek. But I dint.

I combed through the birds like a game of Where's Waldo. "Hang on," I said, when Peter wanted to give me a hint. Each bird looked just like the others. I went from bird to bird, my eyes tracing the contours of rounded deep auburn heads, black bodies. What was I looking for? A different tail length, eye color, leg color? For a while maintaining my meticulous inspection of each bird felt fun, but soon I became tired of the cowbirds.

"OK, I give up."

"See how that one has a longer tail and a thinner bill?"

I focused, straining to take in these differences. "I guess," I said.

"Shiny Cowbird," he said. "Belongs in Mexico."

The Shiny arrived in North America in 1985, then extended north, arriving in the Everglades in 1987. They have even been documented as far north as Canada.

"You woke me up for this?" I asked. "Let's go find that Flamingo." Me, I wanted a big, pink bird.

Our trip had started in Miami; the first Florida bird was a Boat-tailed Grackle, a big, iridescent black bird with a long tail. This "first bird of a trip" had become one of my favorites in the long list of bird games, though it usually ends up being a Rock Dove—known more often and less poetically as a pigeon—floating about the airport. We drove north to Port St. Lucie, where Peter's sister Mary and brother-in-law own a winter home. Mary, who had welcomed us at New Year's, now greeted us once again.

"Welcome Su-Su." Mary gives everyone a nickname, and mine was that of Peter's grade-school sweetheart. I could picture Peter, the red-headed kid, with his quiet awkward young feelings for Su-Su. I loved imagining them holding hands, recapturing the simple thrill of those early crushes.

Mary's house was unnervingly neat yet still welcoming, the guest room stuffed with colorful pillows. Mary had left a basket of gifts for me on the bed, including a mug with the image of a Cardinal etched on the side.

"Bird junk," Peter said as he unpacked.

"It's a great mug," I countered, the weight and size satisfying. Tasteful bird junk. Still, I got what Peter was saying: the birding world is filled with a kitschy stream of bird-themed T-shirts, caps, coasters, and cards—the sorts of things that went well with a needlepoint "home sweet home" over the front door. Yet I was pleased that I had now reached that point in my birding life where people gave me bird-themed gifts.

"You have to let people know you hate this shit," Peter said. He had one green T-shirt that said, "gone pishing." Everything else bird-related he threw away.

"You want me to tell your sister I hate these gifts?"

"Mary? You can tell her anything."

In many ways Mary reminded me of myself. When we arrived she was stretched out on the couch reading a book and had been spending long solitary days at home while her husband traveled for work. With our arrival she became gregarious, quick to tell a joke, ask questions, and to laugh. Maybe that was why I had the peculiar sense Peter and I were a brother-sister duo romping down the trail in pursuit of birds.

Would I have appreciated Mary telling me she hated gifts I had given her? I probably would have laughed. But the truth is, I didn't hate the gifts. I appreciated the welcome that had inspired them, and how immediately and warmly she and the rest of the family had taken me in. I had never experienced this in a relationship before, most of my lovers distanced from, or disowned by, their families.

As we joined Mary on a walk, I slipped my arm through hers.

"I love my mug," I said.

"Don't let Peter throw it away," she laughed.

Soon we reached a swampy area, where we greeted the Tricolored Heron, slim and gangly, and Little Blue Heron whose compact body looked like a ripe fig. Snowy Egrets stepped near the trail with their black legs and yellow feet. Were they golden slippers? Hip basketball shoes? The ethereal Limpkin, which sparkled with diamond-laced wings, hypnotized me. What fun it was being in a land of long-legged birds. And on every post seemed to squat a brooding Anhinga, the water turkey of the south.

At dusk while Mary and I talked as we prepared a salad laced with avocado and hard-boiled eggs for dinner, Peter scooted out in pursuit of rails in a nearby marsh. Even with his own family, Peter did not ease up on his birding timetable.

"Let's go," he said after we had been there one day.

I thought it rude to leave so quickly.

"Listen, we've never been on a proper birding trip."

Really? To me, every trip, even Paris, had been a birding trip. For Peter a proper birding trip had a rigorous definition. It entailed going to a destination—as he had to Brownsville, Texas, *before he met me*—and only birding. No friends, no socializing, no family. On this trip, forget about the bathing suit; there would be no swimming in that beautiful ocean. The drill: up, bird, to bed. In fact, to my horror, even having a decent meal might be breaking the code of a proper birding trip. Anything beyond a goodnight kiss: *streng verboten.* Sleep mattered. Was this really how birders conducted their lives? I didn't want to separate birding from the rest of life; I wanted the birds to be there on family vacations and walks with friends. I wanted to eat well! For now, though, Peter won out. We left on our proper birding trip.

We set our course for the Everglades, and once there I set my heart on seeing a Flamingo. For a chance at a Flamingo we had to walk the flat 1.7 miles out Snake Bight Trail. In a straight shot south, it led from the road to a bight, a bay within a bay.

"Think of all the people who walked down this path," Peter said as we headed off down the trail.

"Who?"

"Pretty much anyone in North America who wants to see a Flamingo."

From the start of my birding life I had reveled in the history, whether it was John Burroughs on Slide Mountain or Florence Merriam Bailey looking through her opera glass. Now here I was in a historical hotspot.

Burroughs and Frank Chapman visited the Everglades, and no doubt walked this path. Roger Tory Peterson and his British friend James Fisher in 1953 forged down this path on their one-hundred-day, thirty-thousand-mile tour of the United States, which resulted in the classic *Wild America*. Beyond these famous birders, thousands of people like Peter and me, who just wanted a glimpse of a Flamingo, had walked this path. Hearing that echo of the past energized me, as if we were on a birder's pilgrimage.

But what an unfortunate *camino*. The sun shone down—beat down ("glade" in Anglo-Saxon means "shining" or "bright")—through the stout mangrove trees that rimmed Snake Bight Trail. I stepped with caution, alert for snakes, yet only a few brown watersnakes slid off into the mangroves. Fifteen feet into the brush a canal half filled with murky water paralleled the trail, a perfect breeding ground for mosquitoes. I strained to register the charms of this flat land.

At the end of the path a boardwalk extended into Snake Bight, a small bay where at low tide hundreds of birds waded in the distance. Was there a Flamingo among them?

We squinted and scoped and hoped for two hours until thirst, hunger, and tired legs demanded we stop. Why had we not brought food and water? On the return we slapped mosquitoes, also regretting we had not brought bug repellant.

As we neared our car, branches snapped and crunched from the dense trailside brush. We both stopped, curious what clumsy animal could be making such a distinct noise. Soon, an alligator long-legged its way onto the strip of cut grass in front of our car. It paused, its long tail thickly menacing, its head slightly cocked.

"Oh," I said a bit stunned.

I was still adjusting to the fact that alligators were everywhere in Florida. Mostly they snoozed, heaved up on land, tails draped in the water, spikily serene, their jagged teeth innocuous. Looking down from a wooden boardwalk, we had cooed over a cluster of baby alligators, just barely a foot long, swimming together in a tangled mass in the marsh below. Like all baby animals they looked sweet. An adult alligator, several feet away, humpfing about, was an entirely different matter. I had read stories of how alligators grab prey and spin, twisting and turning—for instance a leg of a middle-aged woman—until the prey is drowned or worn down, bleeding slowly to death. We backed up as the alligator settled onto the grass, making a low, guttural hiss.

"That's a good bird," Peter offered.

So often our best bird of the day wasn't a bird at all.

In Roger Tory Peterson's *Wild America*, he describes walking Snake Bight, noting that it is just east of "the mangrove key named after Guy Bradley, the Audubon warden who was murdered by the plume hunter Walter Smith in 1905, and who is buried at Cape Sable." Peterson says no more, not about Bradley or Smith or the circumstances of this murder. Still, I didn't doubt it. The Everglades, with murky waters and dense mangroves, gave the sense that bad things had happened, could still happen. And Peterson's line made me want to know more about this murder.

Guy Bradley was a plume hunter. He grew up in Flamingo at the end of the nineteenth century, then a place where people disappeared or reinvented themselves while trying to make their fortunes: in plumes, in land, or bootlegging what a *New York Times* article referred to as "a satanic brew known as Cape Sable augerdent, which could loosen teeth at 10 paces."

Young Guy was not a good shot. He compensated by learning the plume business from one of the best, a man named the Old Frenchman. The Old Frenchman had a market in Paris making "fifty cents for pelican skins, twenty-five cents for sea swallows and least terns, ten dollars for great white herons and twenty-five dollars for flamingo skins." That was a lot of incentive to shoot birds, and there were a lot of birds to shoot.

Florida in the nineteenth century overflowed with birds, was "a perfect cloud of birds." Into this world of magnificent abundance the plume hunters shoved their guns. It is easy to kill a bird sitting on a nest, tending its young. That is what the hunters did, leaving the young to starve to death in the Florida sun, or to be eaten by raccoons or vultures. In little over a decade many rookeries had been shot up. This massacre had only one purpose: to gather the plumes that adorned women's hats.

Stories of the bird carnage traveled north, to women like Florence Merriam Bailey who worked to spread word of the horror of the bird slaughter to women who wore hats with plumes. If the fashion could just be changed, the birds might be spared. Taking a different tack, William Dutcher, chair of the AOU's newly formed bird protection committee, believed that the problem lay not with the plume hunters or with the women in their hats but with the milliners, who exploited at both ends. He decided on education and legislation in order to stop the trade. Dutcher worked with Florida representatives to pass "An Act for the Protection of Birds and Their Nests and Eggs, and Prescribing a Penalty for any Violation Thereof" in May 1901. Dutcher now needed someone to enforce this act, someone who knew the Everglades. He hired Guy Bradley.

Bradley's first year as warden, 1903, was not a success. At thirty-two dollars the ounce, the price paid for egret plumes brought twice that paid for gold. Traveling through the glades to the many and widely scattered rookeries proved so difficult that one man could not keep the greedy plume hunters from the birds. Among the disasters that year was that Cuthbert, the most prized, the most spectacular rookery, was "shot out." "You could've walked right around the rookery on them birds' bodies, between four and five hundred of them," Bradley confessed.

The men Bradley arrested for plume hunting were not strangers to him; they were friends and neighbors. The Civil War veteran Walter Smith didn't appreciate it the first time Bradley arrested his son Tom for shooting birds. The next time Bradley tried to arrest Tom, Smith raised his gun and fired. Guy Bradley tumbled into the bottom of his boat where he was found a few days later, floating about in the bay. With Bradley's death, news of this lawless world hit the press, galvanizing the bird conservation movement; with his death, we had our first martyr for bird conservation.

It's awful to think that a single human has to die in order for the plight of birds—who at the peak of plume gathering in this country were being killed by the millions—to come to public attention. Shouldn't the horror of the bird slaughter suffice? Apparently not. And so Guy Bradley was sacrificed for the cause. His death should not have been necessary, but his death was not in vain.

Determined to visit Snake Bight at high tide, which would bring the flamingos in closer, Peter and I once again trekked out Snake Bight Trail. This time we had water, cheese, apples, and bug spray so we could settle in to wait. At the viewing platform, we set up our scope, picking through the distant birds that peppered the smooth, wide bay.

Big white birds glinted in the sun, tacking across the sky like sailboats in a steady breeze. White Ibis. Great Egret. White Pelicans. From time to time one of us would call, "Pink one, heading right." Excited pause. "Roseate." It felt wrong to dismiss the Roseate Spoonbills with their goofy spatula bills, a fascinating bird in any other context. But we had only one pink goal, the Flamingo.

Minutes, then an hour passed in the heat. I strained to make out shape and color in the sharp sun, my eyes burning, my lower back crying out. When I tired of dots in the distance, I entertained myself with a White Ibis that tiptoed near the shoreline. It's not the most attractive bird with its sky blue eye set in a featherless pink-tinted face. It comes off as sorrowful. Yet I found myself fascinated by its thick, curved bill that gave it a unique seriousness. Above all, it belonged to the royalty of long-legged birds, which made up such a wonderful part of this trip. When the Ibis wandered off I turned my attention to a Brown Pelican docked on a stump like a plump, coy maiden, winking its blue eyes. It seemed very aware of my gaze—perhaps hoping for a handout—but disguised its gluttony with shy glances as it tucked its long bill against its throat, wrinkled like an accordion. How had the ornithologist Edward Forbush dared to describe this bird as "grotesque-looking?" It was nothing but enchanting. Though I wanted to see a Flamingo, I realized I was happy with the birds that were making themselves seen.

"What if we walked out toward the birds?" I suggested. It was an unsavory proposition as a rotting stench rose from the swampy land.

Peter poked along the water's edge. "You know what this smells like?" He was sinking fast into the muck. "When you boil a deer's head."

Peter's comment reminded me of his hunting past, that he grew up shooting birds, squirrels, deer. Had he really boiled a deer head? It struck me as odd, but obvious, that his ability with a gun related to his talent as a birder. He knew how to see movement, shape in the woods. Could he have been a plume hunter? Born today, might the plume hunters have been birders?

I wondered what the Everglades would look like if I let the smell rising from the bay take over my vision. Somehow the national park designation led me to see aspects of this land as beautiful, but if I shed those preservationist sunglasses, I saw a miasmic mess: flat land, still and murky water that encouraged mosquitos to breed, and a tangle of plant life. This was not a landscape I am drawn to, the walks limited by the few short trails carved through the dense brush or on elevated walkways. It took some imagination to see the beauty.

The one who transformed how we all see this land is Marjory Stoneman Douglas. Douglas could see how unique this land was—"there are no other Everglades in the world"—and was able to write about it vividly enough to make people pay attention. In 1947 she published *The Everglades: A River of Grass* as part of the Rivers of America series. Her words helped to convert this land from treacherous swamp to natural treasure. In December of the same year, the federal government acknowledged this natural treasure, making it a national park.

Douglas was an early crusader for the environment. "It is a woman's business to be interested in the environment," she writes. "It's an extended form of housekeeping." I'm not sure I appreciate women being relegated to housekeepers, and yet if the understanding of housekeeping is appreciating beauty in things having their place, then it's the perfect word. And the truth is I love certain kinds of housekeeping, like picking up garbage.

Every spring, my friend Emily and I organize a sweep of the Tivoli Bays, getting locals as well as students out in canoes to haul Styrofoam and old tires out of the water. When the clean-up crews first arrive, the Bays appear clean. But once eyes adjust to seek out foreign objects, out of the muck emerges a stray flip flop wedged at the base of a stand of phragmites. With my garbage-seeking goggles on, I see more and more: engine oil containers and folding chairs, and once a small refrigerator.

Finding the bits of garbage in the Bays is not unlike finding birds. Once you know how to look, there are more birds. The world is richer in garbage than I ever imagined. And then, just as with birding, because I am straining to see one thing I see others: one spring on clean-up day I encountered six Least Bitterns among the cattails.

My metal canoe always wobbles as I paddle to shore with my loot—and I think of it as loot—piled high. All who have been out for hours combing the Bays compare who found the best piece of trash. Is it the eyeless doll or the human-sized piece of pink Styrofoam or the blue plastic barrel? It's all great in my mind, a strange garbage-gathering satisfaction taking over. I stand on the wooden dock looking out into the North Bay, and even though it doesn't look much different, I know it's cleaner.

So maybe Douglas was right—women are the great housekeepers of the world (even if some, like me, may not have entirely tidy homes). If it begins with housekeeping, women's work for the environment extends far beyond. In the bird conservation world, there were many early crusaders, like Harriet Hemenway and Minna Hall, who together started the Massachusetts Audubon Society, the seed for the National Audubon Society; or Florence Merriam Bailey and Olive Thorne Miller, who both wrote about birds and bird protection; or Rosalie Edge who created Hawk Mountain Sanctuary in Pennsylvania to protect the migrating birds from hunters. Would these women hold a greater place in our bird story if the Audubon Society were named the (Florence Merriam) Bailey Society? Would we value house cleaning more if men did it?

In 80 CE the Roman poet Marcus Valerius Martialis, known to most as Martial, the king of the epigram, wrote: "My red wing gives me my name; but it is my tongue that is considered savoury by epicures. What, if my tongue had been able to sing?"

The red bird is the Flamingo; Flamingo tongue was a delicacy.

That a bird has a tongue is not something I think about. I look at the bill of the bird that shreds flesh or cracks nuts. What hides inside that bill is not seen unless a photograph captures the bird agape, like a gull screaming into the wind. Then we get a glimpse of a thin, pointed tongue.

A flamingo, though, has a large, soft, fleshy tongue. Images of flamingos feeding show them head down in the water. They are not capturing fish in this awkward position; they are working water through filters called lamellae that function like baleen in whales. They move those fat tongues back and forth drawing water through the filters.

Hope of seeing these fascinating pink birds kept Peter and me standing in the too-hot sun for hours. Just as I reached for the scope to fold up the legs to begin the flat walk back to the car, the unmistakable, grinding cackle of a rail emerged from the reeds. The bird poked its head out of the reeds, then emerged, a squat, small chicken-like bird. It scurried from one tuft of grass to another. Then a relative on the mainland hollered back. Soon, we had a spontaneous rail fest. I watched, giddy, as the birds with thick down-turned bills erratically ran about, short tails cocked. They had buffy, barred flanks and brownish chests. From time to time they slipped into the reeds and, thin as a rail, vanished like the best of magic tricks. Only to start cackling again, then racing out as if goosed into the open.

Were they King or Clapper Rails? The birds are similar enough that Peter wrestled with the ID into the evening, into the next day, then the next week. He enlarged his photos to get greater detail, examined the images in the guide, decided one minute on King and an hour later on Clapper. Did it have a dark chest or not? Could we match the voice we heard to the tapes? Peter's determination to figure it out and also that he was stumped pleased me—I liked that he didn't know when most often he was so confident.

I now understood the beauty of being able to name birds, and that naming was the first step toward knowing. But the name wasn't the key, the open sesame to my heart. The bird was that key, and in this moment, I was willing to let naming go. I had watched the birds scurrying about their rail lives in that smelly swamp; that was enough.

On this trip, Peter and I had devoted hours to finding a Flamingo. I might have felt disappointed that luck had eluded us. But I did not. It wasn't that the other birds, like the Snail Kites, elegant in flight, or the elusive Mangrove Cuckoo, which flew into a tree not twenty feet from us, made up for this miss. It's that not getting what I desired seemed right, a way that the natural world, the birds, remained elusive, mysterious.

The last night of our proper birding trip to the Everglades, we set up our tent in the Pine Key Campground. Through the night we listened to the Chuck-will's-widow song. I hoped to see the bird, with its goofy whiskers, called rictal bristles. But in the dark, we couldn't make out even its silhouette, which is the shape of an owl laid on its side. The rich calls from deep within the pine were not, to my ear, saying *Chuck-will's-widow*. Rather, I imagined the insistent *Chip-wido-wido* to be the ghosts from the past, the ghost of Guy Bradley still protecting the birds in this strange, beautiful land.

LITTLE BROTHER HENSLOW

Ames, New York

There's a wonderful *Doonesbury* cartoon that I read in a newspaper from 1982, where Dick, wearing a wide-brimmed hat and bow tie, and Thad, in a checkered beret, are birding Matagorda Island in Texas.

"We're in luck, Thad," Dick says. "Matilda's still here, and look, there's Sampson and Walter!"

"I don't know how you keep them all straight, Dick," Thad responds.

"Well, there are only 73 Whooping Cranes on Matagorda, Thad. Besides, they're as individual as you and I. Color, markings, behavior patterns, each bird has its own eccentricities. For instance, Sam there is gay, and Tilly won't eat water beetles unless they've been strained, and Walter spends a lot of time with the ducks."

"And all this is condoned by the wildlife service?"

"It's a sanctuary, Thad."

"I think I'm beginning to understand the Whooping Crane's brush with extinction."

It's a subtle punch line, but Gary Trudeau, the author of *Doonesbury*, got the bird problem right: fussy birds suffer. You can't want your water beetles strained in a world where water beetles are vanishing. You have to take your beetles as they come and be prepared to move on to worms. The Whooper, however, does not have an eating issue, as it is omnivorous. One of the biggest challenges to Whooper populations comes from habitat quality, on both breeding and wintering grounds. Even if Trudeau got the Whooper problem wrong he got something else right. Dick names his seventy-three Whoopers, speaks of them as individuals, as if they are friends. He understands that birds are just like us: queer, complicated, and riddled with quirks.

This sense that birds are just like us has a long history. Harriet Mann Miller, born in Auburn, New York, in 1831, wrote about birds under the pen name Olive Thorne Miller. In her first bird book, *Bird Ways*, she writes: "A well-known French man of letters wrote a book, nearly thirty years ago, with the express object to 'reveal the bird as soul, to show that it is a person.'" That Frenchman was the nineteenth-century historian Jules Michelet, who penned a nearly unreadable (in both English and French) work, *The Bird*. Somehow Miller managed to wade through his prose emerging energized in the fight for bird protection. She felt his words would help to "diminish the enormous slaughter for purposes of personal adornment, of ministering to our appetites, adding to our collections, or, worst of all, gratifying our love of murder, pure and simple." Then Olive Thorne Miller really got going—*allons enfants!*—"Throw aside your prejudice, your traditional and derived opinions. Dismiss your pride, and acknowledge a kindred in which there is nothing to make one ashamed. What are these? They are your brothers."

The birds are our brothers?

I don't search for birds hoping to see a brother or sister. To search for a sibling is to seek the self, and what I appreciate most about birds is that they are, with their winged, straightforward lives, so fully not me. I see birds because I want to see birds. Having them near is always invigorating, unlike people who are often complicated in exhausting ways. Let's not corrupt the beauty of a bird's life by imagining that birds are like us.

Miller, who did not start birding until she was fifty, took her own thoughts to heart, living with birds as if with family. She described buying a Hermit Thrush "at an insultingly low price" and a Wood Thrush from her local pet store. Miller kept the birds in cages, with the doors open, allowing them to fly about her house. She would sit with mirrors on her desk, back to the open cages. When "a shy forester, thinking himself unobserved, said or did anything interesting, [she] would quietly reach for the notebook labeled with his name, and put it all down in black and white." These bird dossiers remind me of new parents reporting on a child's latest accomplishments: He can walk! She said mama!

Seen this way, Miller was that crazy bird lady, the peculiar aunt you tell stories about and don't entirely look forward to visiting. And yet her life was not without human companionship. She was married to Watts Todd Miller and had four children, whom I assume she tended to (almost) as well as she tended to the birds.

One of the gifts of being in such close proximity with her feathered family was that she was able to observe them in detail. Her Wood Thrush: "Nothing can be more attractive than the soft cinnamon browns of his back and wings, and the satiny white of breast and under parts, tinged in places with buff, and decorated profusely with lance-shaped spots of brown." When the Wood Thrush sang, Miller noted, it embodied the spirit of freedom, the song evidence of its "untroubled soul." This intelligent bird learns his name and comes when he is called. These detailed, individual observations of birds that have such human characteristics became *Bird Ways*, published in 1885. This was the first of eleven bird-related books that Miller authored—all after the age of fifty. Her story gave me courage. If Miller could learn her birds *and* write about them starting in late middle age, my dream of learning the birds was not such a long shot. I knew I would not follow her in publishing eleven books—she was a publishing machine, authoring 780 articles in her lifetime as well as twenty-four books.

A photo of Miller appears as a frontispiece at the opening of *Bird Ways*. Her hair is pulled back into a flat bun; a decorative ring of what looks like pearls rests on top of the bun. The curve of her mouth forms a subtle frown, echoing the angle set by the curve of her eyebrows; her eyes gaze into the distance. This dour-looking woman was, however, surprisingly funny. Take, for instance, what she wrote about the Western Wood Pewee in *A Bird Lover in the West*. "A Pewee chorus is a droll and dismal affair. The poor things do their best, no doubt, and they cannot prevent the pessimistic effect it has upon us." By the evening, the repetitive "do, mi, mi, do," song of the pewee gets "lower, sadder, more deliberate, till one feels like running out and committing suicide or annihilating the bird of ill-omen." So though Miller loved her birds, she, like an honest mother, recognized their failings as well as their gifts.

Because Miller was intent on saving birds, whether by bringing them into her home or by writing about them, her descriptions of birds that have a soul, intelligence, and emotion were surely calculated. If the birds are like us—should we not want to protect our own?

The current method of eliciting concern about birds, and above all encouraging people to open wallets to demonstrate that care, is not to ask us to meditate on the bird's feathery soul. Bird advocates pummel us with numbers. For instance, the Henslow's Sparrow population has had a 95 percent decline since the midsixties.

That number does make my stomach drop. But does that motivate me to rush out to save my brothers and sisters? Not really, because the numbers in relation to birds are sadly big, inexcusably endless, and impersonal. For me to care, for me to send my dutiful checks to the Cornell Lab of Ornithology, to Audubon, and to other conservation organizations, for me to call legislators or show up in the field to volunteer, I need to meet the birds, to look them in the eye and hear them sing.

On a hot July day, Peter and I packed up the car with snacks and water and pointed north, tempted by a report of an Upland Sandpiper. More properly a grasspiper, these birds live in the uplands, areas that are dry, open grasslands. Beyond hoping for the Uppy, we were meandering, exploring a farming landscape that seemed to exist out of time, west of Albany and south of route 90. On the lazy, curving roads, we spied Kestrels perched on the phone lines and Red-tailed Hawks contemplating murder from atop electric poles. I liked birding the quiet of July, having each song distinct; after the rush of spring migration it felt like moving from the symphony to a sweet quartet, where we were an audience of two.

On West Ames Road we scanned the rolling fields, some plowed, some with corn almost knee high, and some untended, filled with wild parsnip, like a green version of Queen Anne's lace. I drove, windows down, the breeze gentle, and at thirty miles an hour Peter picked up sounds that erupted from the wide fields. Bobolink. Savannah Sparrow. Meadowlark.

"Stop," Peter directed. He heard the Uppy, which sings a song that is hard to mistake. I pulled the car to the shoulder. The Upland Sandpiper called again, a crazy, long "mournful but mellow, rolling whistle like that of autumn wind, 'wh-e-e-e-e-e-e-e-e-e-e-o-o-o-o-o-o-o-o.'" We found the long-legged shorebird working a field on the north side of the road. The tall grass camouflaged its speckled breast and long straight bill. With our scope, we spotted two downy young scurrying next to the mother.

We soon folded back into the car, deciding to find a nearby place for the night. As I drove toward Sharon Springs, I could feel the tightness in the sky as the day arced toward end. A narrow country road that flowed uphill between open fields drew me in.

"Mind if I see what's up here?" A sort of gambler's hunch tugged me down the road. Then again, it could be that narrow, quiet, rolling roads that lead nowhere and everywhere are seductive. Earlier, we had passed Amish-driven buggies along these roads.

Though I say I don't gamble, I also know that we gamble in a way all the time. I pick this job and not that one, speculating that it will make me happy, productive, support me. Mid-age, these choices take on greater weight. I chose to learn about birds, so I had given up the dream of learning Spanish, the hope of finishing the wooden boat I had started to build before the birds swept me away, the notion I had that I would trek up a peak in Nepal. Had I bet right on Peter? He was a perfect companion on a weekend of exploring, our energy to venture well matched. After our back and forth earlier in the year, we had reached a calm compatibility. Most of the time we were like a pair of Ring-necked Ducks puddling about, cheerfully inattentive yet never far from each other. Maybe at my age this was the best I could hope for: admiration, contentedness, kindness, and affection. For my part, I felt like a lucky gambler, that in all the swamps I had looked through, I had found someone who shared my new passion, someone who guided and encouraged my learning.

At the top of the hill, we had long views over the fields and onto the edge of the horizon. The day had been crisp, like sheets dried in the sun. At a level crest on MacPhail Road, a narrow two-lane with no middle line, Peter pointed to the shoulder. "Pull over here a sec."

"I love this sort of landscape," I mused. No spectacular cliffs, soaring peaks, or rushing rivers, just fields, gentle slopes, and untended land. What surrounded us was a sweet emptiness that reminded me of being seven when nothing sad had yet happened in my life.

Peter pushed back into his car seat, tired and relaxed. He reached over, rubbing my shoulder. We sat like that for a while, the silence delicious. I drifted into a liminal state, content like after a good meal. I wondered what it would be like if such a state extended. Perhaps as with too much happiness, it would become intolerable, the texture of life flattened. For the moment, though, I wanted it to go on and on, the silence so deep I could hear my own heartbeat.

My reverie ended with a start as Peter sat up, back stiff as if he'd been yanked forward by an invisible leash.

"My god, Henslow's."

I tilted my head toward the window, while Peter pointed. A faint sound, an airy *tsi-lick* that Peter described as the wonderful-pitiful song of the Henslow's Sparrow, rose from the end of the field. Yet I could have easily failed to let it register, sitting there drifting into daydreams.

We hurdled into the field, waist-high in the grasses. Peter motioned me to stop. One hundred feet in front of us the little bird balanced on a parsnip stalk, singing its dinky song. It had a light brown breast and streaked upper parts. Most distinctive was the heavy bill protruding from a flat-topped head. The bird seemed charmingly misproportioned.

The Henslow's Sparrow doesn't belong in this part of New York State. Its range ends west of us; even there it is, according to an article in the online bible of birds, Birds of North America Online, "uncommon and famously inconspicuous." It's one of those species with fussy needs: damp, grassy meadows that include matted vegetation and a variety of weeds. Those wide, empty, weed-choked fields are less and less common. We mow, we farm, we build houses—we destroy the habitat that the Henslow's needs to breed.

Peter waded amidst the parsnips, camera to his eye, while I stood, binoculars at the ready when the bird emerged. We played this game of up and down, hope and thrill for half an hour before returning to the car. Peter was red-faced and beaming, his pleasure pure and infectious. I clapped as after a beautiful performance and felt in that moment that that little sparrow was the most remarkable creature on earth.

What if I hadn't turned down MacPhail Road? What if Peter hadn't asked me to stop? What if the bird had waited ten minutes before singing, and we were gone? All of the little gambles that led to a bird transfixed me: How could I be so lucky?

Peter's a sparrow man. I knew this from the start, when we got up at dawn in Maine for the Nelson's. Only now, seeing the Henslow's, did I understand what that meant.

Because sparrows are not the sexiest birds in the sky, it takes a different sort of affection to come to know them. To be a sparrow fan is to be committed to finding the secrets in the fields and woods. Only birders love sparrows, and only certain birders love sparrows. Peter finds good sparrows because he looks for them.

Peter does not gravitate to the classic beauties, the catches. He had an eye for that smudge of mustard behind the eye, that bigger beak, that weaker flight. To Peter these details are beautiful. A wash of affection hit me as I realized that Peter championed the underappreciated, the overlooked; Peter loves the less lovable.

Am I less lovable? I doubt anyone likes to think of herself that way. And yet I saw that I wasn't easy to love, hesitant as I was to fully give over to this or any relationship. As much as I loved Peter, I saw that I had one foot dangling out the car door, ready to leap out, even if it meant getting scraped and bruised. "Love like your heart is going to be broken," Peter challenged me early on, knowing perhaps that I loved a challenge and that I would fail. Had I ever allowed myself to fall in love? *Fall.* In love.

Maybe it was the language of love that had made this such a challenge. Had I not spent my life avoiding falling? When I led a climb, I always put in more gear than necessary, lacing up a route like a seamstress making smaller stitches so that her handiwork will hold. And, as I reached upward, I always kept an eye on where I might fall, how long a fall it would be, and what I would hit on the way down. Could I change, only look up, trusting my strength and skill enough to step upward, fear in check, feet solid? I liked to think I could do that, that I could do anything. But the older you get the more you know how hard it is to change how you see the world. And yet my eyes *were* different. It's not just that the optometrist confirmed my vision had improved—exercising the eyes in search of birds had an effect I hadn't anticipated—it was that my eyes now focused *out* to the world.

That effort of looking out, always hopeful to see something, and of listening intently silenced the inner chatter and corralled the wandering thoughts. My outer focus brought on an inner calm that had grown inside of me in the last years. This state of attentive calm felt so vibrant, I called it love. I hadn't fallen in love; with the birds I had walked into love, peered my way to love. And I understood that learning to love, like learning the birds, doesn't happen in a day or week, or in one night of passion. It takes effort, constancy, devotion.

That evening, in a cozy B&B a fifteen-minute drive from Henslow's field, Peter called a few friends to alert them to the Henslow's. Before a spare dinner of cheese and crackers (any open restaurant was miles away, miles we did not want to drive), he posted it to the New York State birding list.

"That good a bird?" I asked. I was still learning the scale of rarities, what birds were special to our region or special to the state.

"For you to see the Henslow." He was brushing his teeth and stopped to spit. "That's like being president before being president of your class."

The Henslow's was exciting enough that for the next week, we learned from the birding list, birders drove up from Manhattan, four and a half hours away, to greet the little sparrow.

We had found the bird because of its simple song, which Sibley describes in his guide as a "feeble hiccup." That seemed an accurate, if heart-breaking, description. Few are impressed with the bird's song. In 1881 an ornithologist reporting in the *Ornithologist and Oölogist* wrote:

> The musical performance of this bird has very little to commend it. When the muse inspires his breast he mounts to the top of a weed or some other object that raises him just above the grass. There he sits demurely until the spirit moves, when he suddenly throws up his head and with an appearance of much effort, jerks out his monosyllabic "tsip" apparently with great satisfaction. Then, having relieved himself he drops his head and waits patiently for his little cup to fill again. Somehow I cannot watch him while thus engaged, without a feeling of pity for a creature so constituted that he can be satisfied with such a performance.

Is this bird high on the conservation list because it is pitiable? Or is it there because it has experienced the steepest decline of any grassland bird, 7.5 percent a year over the past three decades? Maybe a combination of numbers and emotion is needed to urge us to care for our declining species.

Or maybe it's even simpler than that. Until that dreamy day on a desolate road when a bird named Henslow's Sparrow appeared, I had not given the bird a thought, except to note the name when flipping through my guidebook. Now, its song tattooed onto my heart, its fragile legs clutching the parsnip, I cared. If the Henslow's were my brother, I would call and text asking for news: Did they mow your field? Did you find a mate? How are the little ones?

The next morning we went back to visit the Henslow's Sparrow. I have a photograph of Peter sitting in a folding chair I kept in the back of my car. He has on a light blue baseball cap, his red hair licking out at the edges. His hands are cupped around his ears to amplify the sound, should the bird sing. Peter was convinced the bird would not reappear. Still, we waited, the sun raining down, the wide field stretched before us.

According to ornithologist Edward Forbush, it was odd that the bird was not singing. "Some of the males have the habit of singing, if singing it can be called, after dark; sometimes they sing until midnight, and in some cases nearly all night. . . . It seems as if the bird just sings in its sleep." Maybe it had worn itself out, having sung all night and now taking a morning nap.

Needing to stretch my legs, I left Peter to wait on his own while I walked up the hill, past fields enclosed with wooden post-and-rail fencing. That one little bird had chosen a big area to set up shop. How had he found this plot of land, I wondered. Orange daylilies bloomed along the edge of the road. Clouds scuttled across the sky, which curved to the horizon as if revealing the shape of the earth. Might Henslow himself have walked these Henslowian fields?

No. John Stevens Henslow did not know this land; he lived in England, working as a botanist and minister in the early to mid-nineteenth century. Henslow is most celebrated for ceding his place on the *Beagle* to his student, Charles Darwin. That small but gracious gesture changed history. Henslow was also a confidant to Audubon, who first found the bird that had us smitten in 1829 in Kentucky. Audubon named the bird for

Henslow "to manifest my gratitude for the many kind attentions which he has shewn towards me." But really, Audubon could have given the bird a better name—perhaps the night-singing sparrow, or better, a name used by indigenous people who had first lived on this land.

At the top of the hill a plaque tacked to a large boulder informed that *Near this spot Catherine Merckley on October 18, 1780, fleeing on horse-back from the Indians was shot and scalped by Seths Henry.* The violence of this past contrasted with our gentle summer day. There were so many lenses through which to view the shape of this land. Where was the plaque that would tell me more about Seths Henry, chief of the Schoharie Indians. Were there losses to his tribe? How different might this land be if it were still tended by the original inhabitants.

As I walked back down the hill Peter appeared a speck in the distance, now standing.

"The bird is back." I hadn't seen him this happy in a while. His affection for the Henslow's seemed profound, the sort of love one feels when, after long separation, one sees a best friend or favorite sibling.

"Someone was scalped up the hill."

"I think we're safe," he said.

"That was in 1780."

"Would you go ask permission to walk out this field?" Peter requested. In the light of day, it didn't seem wise to openly trespass as we had the evening before.

I didn't enjoy knocking on strangers' doors, but preferred that to worrying someone would stop us on this bird quest. A car sat idle in the driveway of the house across the road. I assumed the homeowner might be connected to the land where the Henslow's had taken up residence. A middle-aged man came to the door in a white t-shirt and jeans, clearly not used to people knocking.

"Hi," I said, as I'm-not-here-to-sell-you-anything-or-convert-you-to-anything friendly as I could be. "Is that your field across the way?"

"Belongs to someone in Ames, name of," he hesitated. "I think his name is Jones."

"Do you think we could walk out on it?"

He peered over my shoulder.

I held up my binoculars. "We're looking for birds. Turns out there's a special bird in the field."

"An eagle?"

Everyone wants a special bird to be a big bird, a majestic bird. It was hard breaking it to him that we were excited about a blob of a brown thing. His eyes wandered as I tried to transmit the importance of the Henslow's Sparrow.

"I don't think anyone will care," he said. I wasn't sure if he meant no one would care about the bird or about the land, but I took that as a yes.

"We have permission," I called.

Peter took off into the field, waist high in the grass, eager to capture images of his little brother Henslow singing its sparrow heart out in the grasses outside of Ames, New York.

INTERLUDE

The Other Leopold

Tivoli and New York City, New York

"He loved birds. Leopold."

"Aldo? He was a tree man. Not a bird man."

"No, not Aldo. The other Leopold."

For me, there is only one Leopold, Aldo, the Midwestern environmentalist who wrote *Sand County Almanac*.

"Nathan."

"Nathan Leopold?" This does not make sense.

"He loved birds."

"Leopold of Leopold and Loeb? The murderer?"

"Yes, the murderer."

In this way, on a quiet late fall evening, I learned that Nathan Leopold, famous for teaming up with Richard Loeb to commit the crime of the century by murdering fourteen-year-old Bobby Franks, was a *birder*.

The voice on the other end of the phone was the president of the small college where I teach. Leon, as we all call him, knows nothing about birds.

"Why do you know this?"

Leon had just had dinner with the Nobel prize–winning scientist James Watson, who codiscovered the double helix structure of DNA. Now eighty-seven years old, Watson had just published his memoir, which is really a tribute to his father. His father was a birder; Watson birded with Leopold.

"Read Watson's memoir," Leon urged, "you'll find it interesting."

These bits of birdy information float toward me, spicing the day like the birds themselves. Like Mary's gift of the bird mug, this sort of bird-related information makes me feel I've slipped on, like a perfect coat, a new identity: I am the bird woman. For the first time in my life, I have the sense that who I am, who I think I am, and how people see me synchronize. Like a juggler, I'm able to catch all three balls and send them in a perfect arc into the air.

Becoming the bird woman happened so quickly—I was but a year and a half into my bird life—that at times I was surprised when, on my morning walk, a neighbor stopped me to ask what bird was killing the birds at her feeder (most likely a Cooper's Hawk, I offered) and a student sent me an email gushing about a bird sighting ("It was so shiny, speckled"; I refer her to the European Starling). Friends sent regular gifts by email or phone: Have you read Robert Frost's poem, "Ovenbird"? or Theodore Roethke's "The Far Field" (*For to come upon warblers in early May / Was to forget time and death*)? At a party, a colleague was stunned I didn't know the Mel Brooks movie *The Producers*. "I'll send you the link," she said, "you'll love it." In the film, Max Bialystock is looking for the "Kraut" Franz Liebkind. "He's up on the roof with his boids. He keeps boids. Dirty . . . disgusting . . . filthy . . . lice-ridden boids."

And then this golden birdy tidbit: Leon calling to tell me that Leopold, whose story I know of mostly through films like Hitchcock's *Rope* and the more recent *Swoon*, loved disgusting, filthy, lice-ridden boids.

That Leon learned of Leopold through Watson only added to the shimmer of the story. So I read Watson's fascinating memoir to learn that his father's meeting with Leopold was pretty ordinary: James Watson Sr., just back from the western front in World War I, was birding in Jackson Park in Chicago when he ran into Nathan Leopold. When birders meet in the field, we tend to stop and share sightings, talk bird talk, beginning with: See anything good? Maybe Leopold bragged about his collection of birds—three thousand by the age of fifteen. From that first encounter, the two young men became regular birding partners.

A birding partner is a particular relationship. In many ways it looks like a domestic partnership. You spend hours together, often in silence; you know what they eat, how often they pee; you know how they respond to luck, both bad and good, in finding a bird. You see their ability to focus, and you watch their memory in action as they identify a bird. You are witness to their kindness toward the birds, or their selfishness. You know your birding partner well, and yet also not at all. That makes it a unique relationship: all that you know is filtered through the birds and the bird experience. Often you don't know what they do when they go home, how they make a living, or whom they love.

Did Watson, in those fresh days of scouring the ponds and woods of Chicago, ever have a hint that bright and charming Leopold, half in love with his beautiful friend Richard Loeb, was capable of kidnapping Bobby Franks and bludgeoning the boy to death—arrogantly believing they could commit the perfect crime, were so smart that they were superior to the law? Brilliant as he was, Watson probably did not imagine such things.

Leopold was smart by many ways of measuring smartness: he spoke nine languages, had an IQ of 200, graduated from the University of Chicago at age eighteen. I thought, as I read Watson's memoir, of how at the beginning of my bird journey I translated the two hundred local birds to French verbs. I still struggled with my verbs. If Leopold spoke nine languages, that was thousands of verbs; how awe-inspiring it was to imagine that Leopold might have held in his head all of the ten thousand bird species that live on this planet.

I thought of birders as smart, people with sharp minds, who savored small details. Birders are not necessarily articulate, or well read, and many are socially awkward. But all birders share one characteristic that I associate with intelligence: curiosity. Without curiosity there are no birds. In a perfect loop, curiosity lures the birder out and the birding in turn builds an even greater curiosity. Curiosity is a tonic, one that makes me buoyant, light-headed, happy. I also saw that many (often the best) birders shared another characteristic with Leopold: a confidence that bordered on arrogance.

Leopold was not a casual birder. His fascination with birds had a scientific inquisitiveness, and he was perhaps headed toward a career as an ornithologist. In 1920 Watson and Leopold, along with birder George Porter Lewis, published an eighteen-page pamphlet titled *Spring Migration Notes of the Chicago Area*. Leopold kept a keen eye on migration patterns around Chicago. In 1922, at the age of eighteen, he published his first paper, "Reason and Instinct in Bird Migration," in the prestigious ornithological journal *The Auk*. In this article he looked at a few instances of accidental birds, like the Harris's Sparrow, that then appear more frequently. If migration is controlled by generations of instinct, these changes in a bird's range come about through reason or learning, he concluded. Here was the brilliant young man arriving at conclusions that highlighted the intelligence of birds.

Leopold spent a brief stint of his college years at the University of Michigan, where he studied birds under Norman Asa Wood, famous for finding the first nest of a Kirtland's Warbler. The Kirtland's is one of North America's rarest songbirds, with a very narrow range, breeding in Ontario and in northern Wisconsin but principally in a few counties in northern Michigan, approximately a three-hour drive from where Leopold was in school in Ann Arbor. This is jack pine region and so locals call the Kirtland's jack-pine birds (a name we should all adopt). It is a larger warbler, with a yellow chest, adorned with a breast band made of dark spots that could be read like the dot-dash of Morse code. Wood describes the birds as quick and restless, with a direct, slightly undulating flight. It is named for a renowned Ohio naturalist, Jared Kirtland, on whose farm in Ohio, along the shores of Lake Erie, the first specimen was discovered by his son-in-law, shot, and then classified in 1851.

In 1922, with Watson at his side, Nathan Leopold embarked on a trip to find a Kirtland's nest. That first year, the two young men traveled "well-nigh impassable roads" to the banks of the Au Sable River in Oscoda County. They found no Kirtland's, probably because the trees in that region were too tall; the Kirtland's is an exceptionally fussy bird, preferring trees between five and twelve feet in height. This requirement of tree height is one of many reasons these birds are so challenged. The spring of 1923 the ornithologists journeyed out again, this time for a week of observation and hunting. They left Chicago accompanied by two other young bird researchers, Sydney Stein and Henry Steele. The group settled into the Van Ettan Lodge and immediately set forth to find the warbler. As they drove down a sandy road, there was a song none of them recognized. They got out of the car to track down the bird. After fighting through several hundred yards of dense jack pine, there was the Kirtland's, "every muscle in his body tense," as it let "out a burst of clear, bubbling song." The song was so loud it could be heard a quarter of a mile away. The Kirtland's was a fine adult male in "full nuptial plumage." They watched the bird for a while before heading on to spot other birds and hoping to locate a nest. Only on the next day did they find that much-wanted nest; this was the fourth nest ever found of this secretive warbler.

Watson and Leopold determined to observe the nest. On June 19, from 9:50 to 11:30 a.m., they noted activity every few minutes, including the number of times the bird sang in a minute, how often it flew to the nest, and how agitated it was by their presence. Leopold's notes reveal his mind: careful, thorough, attentive to details.

To document this find, they wanted to photograph and film the birds. In order to have better light, they decided to cut down some of the surrounding jack pines. It's hard to believe any of them thought this a good idea, and it wasn't. Though the birds were "surprisingly tame," once the trees were down, the female became timid, hesitant to feed the young. When she did come in, a Brown-headed Cowbird baby, dwarfing the Kirtland's, snatched the food. So Leopold took things in hand, removing the nestling Brown-headed Cowbird. Without knowing it, Leopold was initiating one of the conservation measures still used today to help the fragile Kirtland's Warbler population: protecting the chicks from being victims of parasitism.

After a lunch break, the men returned for further observation of the nest. Alarmed by the fact that the parents were not attentive to the young, they decided to feed the birds themselves. Lying near the nest, Leopold was able to feed the nestlings two horseflies. Soon, the adult Kirtland's grasped the pleasure of being hand fed and landed on Leopold's thigh and shoe. It was fed a total of seventeen flies. A black-and-white photo, a frame from a moving film, shows Leopold on his belly, his hand enormous as he pinches a fly between index finger and thumb in offering to the warbler.

I was mesmerized by this image of a man who was capable of murdering another person, gently feeding a bird. The photo speaks to the fact that we all hold within ourselves contradictions. In most of us those contradictions are small—we lean toward kindness, then inch in the direction of cruelty. Leopold vaults. And I realized that though I found people who contain oceans of contradiction fascinating, Leopold's leap from tenderness to senseless cruelty was stomach churning. His commitment had become something perverted, not an attractive or admirable obsession. In committing this murder, Leopold had left the realm of obsession for something else. I was revolted that he had killed because he could, like a person who kicks a dog because they can. Could all obsessions tilt into the appalling? The unacceptable?

Through his time feeding those baby warblers, Leopold arrived at a few general notes on the Kirtland's Warbler. He concluded that the nests are often found near roads (suggesting renaming the bird the Road-side Warbler). He observed that the bird doesn't walk, as previously reported, but hops. And he now understood the Kirtland's diet, which consists mostly of centipedes, worms, and caterpillars. This seems intriguing but scant information given the hours of observation. Leopold's contribution to Kirtland's Warbler knowledge makes me marvel over the volumes of detail available on the eating, breeding, nesting, and flight of so many birds. All of this must have taken lifetimes of observation by people capable of watching birds like monks in meditation.

According to Watson, Leopold collected (that is, shot) a few of the Kirtland's, one sent to the Field Museum in Chicago and another to the Cranbrook Institute in Detroit. Watson was concerned with the collecting, "fearing that the loss of only a few birds might tip this species toward total extinction." Leopold did not include information about collecting the birds in the article he published in 1924, "The Kirtland's Warbler in Its Summer Home," which he first delivered in person at the annual meeting of the American Ornithological Union.

What becomes clear in reading this article is that the qualities of a good birder or ornithologist—keen careful observation and a steely patience—are the same qualities needed to plot a murder. Leopold and Loeb spent months working out the details of their almost perfect crime.

I calculated that to see a Kirtland's Warbler it would take a well-timed spring trip to Michigan. Unless, of course, a Kirtland's were to make a rare and unexpected appearance in Central Park, which it did in the spring of 2018, when I was nine years into my birding life. I happened to be in New York City, visiting a friend in the hospital. Sunday morning, after a night spent in his too-quiet apartment, I walked to the park at Eighty-First and Central Park West with little hope of seeing the bird. Still, I knew that the morning air, as well as the focus of the search, would give me courage to shoulder the weight of hospital air and the grief already building around my heart.

I passed the Diana Ross playground, then followed the path as it curved north. Soon, I realized a dizzying number of birds draped from the trees. I stopped to look at them and caught my breath as I realized that before me flitted Bay-breasted and Cape May Warblers, two species that I never see enough in the Hudson Valley. And the birds were not maddeningly singing at the top of a tree, but rather at or near eye level. These birds foraged in plain sight, while I grinned and watched.

Soon, a slight man, nondescript except for his binoculars, approached me and asked if I wanted to see the Kirtland's. I grinned like we had just made a drug deal.

"It just showed up," he said as he walked me back to a tree. "There," he pointed at a bird hopping from branch to branch.

Not possible, I thought, as I put my binoculars to my eyes.

There it was, a small yellow marvel, happily living where it didn't belong. Near me a group of birders had formed like a paramecium, tentacles of scopes and cameras all pointed toward the bird.

I would like to say that I lingered for hours, appreciating every feather, really savoring this once-in-a-lifetime event. But I was cold, and hungry, and anxious about my friend who, like the Kirtland's, was a rarity. I wanted to rejoin the group of adoring and attentive friends milling about outside his hospital room offering solace and jokes and words of love. We all knew this was the end, and so too did he, at turns frightened and agitated and then so clear-eyed.

A week later my friend was dead. And what of the bird? What was its fate after its unfortunate detour through the East Coast? Did it die there in the park? Did anyone notice?

To lose one Kirtland's seemed a sad event; the bird is so precarious in this world. In 1967, the Endangered Species Conservation Act listed the Kirtland's when evidence showed their populations had crashed from one thousand to four hundred birds. Later, with the Endangered Species Act of 1973, the Kirtland's Warbler Recovery Team formed to protect the jack-pine habitat and to guard the nests from Brown-headed Cowbirds. Their efforts, applied diligently over the decades, have worked. In 2012 there were an estimated four thousand Kirtland's Warblers. Though the bird has been removed from the endangered species list, it remains a species that needs the help of watchful scientists and environmentalists.

The work of the Kirtland's Warbler Recovery Team all points back to the murderer-to-be Leopold. In his article, Leopold offers thoughts on why the Kirtland's is so extremely scarce, sounding an early warning that this was a species at risk. His observations are credited as leading to future conservation measures used to help save this warbler.

What if we remembered Leopold as an early crusader for this special warbler, and not only for his sensationalized crime? Making such a shift would be like saying "Nixon" and thinking *Clean Water Act! Clean Air Act!*, rather than Watergate. But a murder like Leopold committed is not a forgettable or forgivable crime. Still, I felt grateful to know the full range of this man; I would always first associate Nathan Leopold with the Kirtland's Warbler.

On May 17, 1924, Leopold with his friend George Lewis tried to bag some Wilson's Phalaropes, which flew out of Wolf Lake, adjacent to Eggers Woods, in Chicago. A Wilson's Phalarope is a sand piper–looking bird, with a needle-like black bill. Like other phalaropes it moves about in a nervous way, often spinning on a pond and using its lobed toes to stir food to the surface.

Four days later, Eggers Wood would become famous not for sheltering phalaropes but for being the site where the body of Bobby Franks was hidden in a culvert. Leopold and Loeb had taken Bobby, who was a distant cousin of Loeb, into their car, killed him, and then transported the body to the woods. The next day, a man making a shortcut through the park came across the body.

Leopold and Loeb were scrupulous, yet all it takes is one small slip. The central piece of evidence that quickly led the police to Leopold's door was a pair of reading glasses found near the body in the woods. Only three such frames had been sold in the Chicago area: one to a woman, one to a man who was overseas at the time of the murder, and one to Nathan Leopold. That narrowed the pool of suspects significantly.

"I told Captain Wolfe I'd been out there [Eggers Woods] recently. I had even tripped that day, not twenty feet from the place the body was found, when I tried to run in my rubber boots to get a shot at a Wilson's phalarope. . . . A Wilson's phalarope is rare enough in the Chicago area so you don't forget about it when you collect one." In his memoir, Leopold described tripping in the police station to show how the glasses could have spilled out of his pocket.

None other than Clarence Darrow, the best lawyer that money could buy, defended the two boys who pleaded guilty to both kidnapping and murder. Both crimes carried the death penalty. Leopold's father asked Watson to testify for his son. Watson refused, fearing for his job and worried about the repercussions if it became public knowledge that he was a long-time friend of the perpetrator of this brutal crime, which at the time was a media sensation. And, really, what could he have added to the testimony? Only that Leopold was a good, sharp birder and that he had tenderly fed Kirtland's Warbler babies by hand. I'm doubtful that would have swayed the jury.

Darrow's stunning twelve-hour-long closing defense convinced the jury not to hang Leopold, who was only nineteen, or Loeb, who was twenty. His defense helped to alter the course of capital punishment in this country. In 1924, Leopold and Loeb were sentenced to life, plus 99 years.

Life plus 99 years without birds.

And yet birds are never far away, even if you are in prison. Open fields surrounded Stateville Penitentiary, located near Joliet, Illinois. Horned Larks and Vesper Sparrows nested in the field behind the prison, and in a marshy area Killdeer darted about, emitting their hysterical cry. From a back window of his prison cell, Leopold could hear the birds. He gazed out over the wall, watching the "gradual onset of the soft spring twilight. A robin perched on the wall and greeted the coming night with his joyous carol. To me the song of the robin has always been one of the most beautiful sounds in the world—and one of the most nostalgic. It brought back vividly many memories of long ago." I am grateful for Leopold that he had these bird songs.

Part of my sympathy for Leopold lies in the fact that I have taught in Bard's prison program, once a week driving out in the evening to

Woodbourne Correctional Facility, a medium-security prison, to teach men who are incarcerated for much of their lives. In our short breaks and after class, the students approached me and asked for things: an article or book to help them write a paper, more supplies like pens or paper. They weren't asking for me to bring in contraband, or a file to chisel their way out of prison, but the ask always had that desperate edge to it. I hesitated the first few times until I finally understood that so deprived of books or birds, family or freedom, they wanted whatever they could get, finding perhaps some solace in these small items and the focus that they offered. Being pulled out of the self, even if for a bit, might be an antidote to despair. I have no doubt that the bird song brought Leopold that solace.

But really my sympathy for Leopold, in jail for life, emerges because I cannot help but think of him as a fellow birder, feeding those Kirtland's babies. And birders are generous in one particular way: we want others to see and hear what we have seen and heard. We want to share the beauty and wonders of the world.

In learning that Leopold was not just a cold-blooded, arrogant murderer—the sort of beautiful, slimy guy in *Rope*—but a fellow birder, I had to expand who I imagined birders to be. Yes, birders are often nerdy in the sweetest and most maddening ways, but overall there was no neat birder profile. I felt silly that I had initially imagined there was one. I had met or read about birder scientists and poets, musicians with their sharp ears, older women and skinny boys, artists who note color and eye rings, egomaniacs and social misfits, presidents, environmentalists, photographers, and murderers. This range appealed to me—anyone could be a birder—and that this range mirrored that of the birds themselves seemed perfect.

Leopold didn't just listen to the birds. Another inmate, described as a "Mexican working in the Fiber Shop," picked up a fledgling Horned Lark and tamed it. It was grown and "looked like a regular bird instead of a powder puff on legs, the way the youngsters do. It was quite tame." Leopold bought this bird for ten sacks of tobacco. The bird would fly to him when he whistled and had the freedom of Leopold's cell in the evenings (though, I admit, it's odd to read of a prison cell as a place of freedom). In a prison, no bird is going to come to a good end, was my first thought.

And sure enough, the bird, left in the stockroom, lets curiosity get the better of him; he pecks a piece of cheese on a mousetrap.

After this Horned Lark, Leopold acquired a few other Larks as pets, one as a fledgling. He showed great determination with this baby bird as for the first few feedings he had to pry open the bird's beak. He also showed great tenderness: "When full of worms, he'll fluff out his feathers and cuddle up hard against your shoe. It's pretty easy to lose your heart to one of the soft little things." So Leopold did have a soft heart, giving it to the soft little birds.

His favorite pet bird was a Robin he named Bum, who flew about visiting the men, whom Leopold referred to as "cons," in their cells. Bum came when Leopold whistled and loved his raisins. Then, one day, the bird was found in a brown paper sack with its neck broken. "If the fellows in E House could have got hold of the man who killed him, I think they might have wrung his neck." I like that prison justice extended to the birds as well.

Clarence Darrow was aware that Leopold might do more with his love of birds. On September 20, 1924, he wrote to Leopold, "I am ambitious for you to write your bird book." Again on March 9, 1928, "I am still anxious that you should have a chance to write a book about birds." Leopold claimed in his memoir that he never wrote a book on birds. In fact, he did.

In 1958, after spending thirty-three years in prison, Leopold was paroled. This unexpected parole was granted because Leopold had volunteered for war-time malaria experiments. It helped that he was also a model prisoner, offering language classes and developing the prison library. One condition of his parole was that he live in a rural region of Puerto Rico where no one had heard of his crime of the century. Aged fifty-seven, he found work as a medical X-ray technician, married Trudi Feldman Garcia de Quevedo, and wrote his bird book, the *Checklist of the Birds of Puerto Rico.*

GOOD BIRD

Santa Rita Mountains and Bisbee, Arizona

In the midnineties two fearless young cats, one irritating one-hundred-pound German Shepard, my girlfriend Sam, and I packed into the cab of a Toyota pickup truck for a long drive west. We had plastic bins in the back of the truck with all we really cared about: several computers, some clothes, essential books. On top of the bins rested our bikes. We headed south on 81, then, once in Tennessee, slipped onto Route 40 for a ride through Arkansas, Oklahoma, and New Mexico. We had wanted to stop along the way—Graceland in Memphis or White Sands in New Mexico—but the heat was such that we could not leave the kittens or they would roast in the truck. So it was a fast drive from our sweet house on a dirt road outside of Austerlitz, New York, to Tucson, where we were greeted by a fabulous apartment in Armory Park, one of the older neighborhoods in this western city. Gravel replaced grass lawns, and mesquite trees replaced oaks. The apartment was airy and light, with eight-foot ceilings, long windows, and wooden floors. As I breathed in the dry air, I realized that in front of me stretched months, maybe years, in which I would not have to shovel snow or haul firewood.

Midthirties, I had decided to go back to school for my MFA in writing. Sam, poet and teacher, was along for the ride. Sam had quit work teaching developmentally disabled children and was looking for the next best thing. My goal was to teach. As a graduate student I taught composition, animated in those early days by my students' rhetorical analysis and persuasive essays. From time to time I led a section of the undergraduate nonfiction workshop. After years of working as an editor I reveled in having my young writers there in front of me, eager to learn. A teaching prize, then a fellowship, affirmed my sense this new life was right for me.

On weekends I hiked every trail around Tucson, swooning over the sense of space and the long views that tipped at the horizon. I admired the goofy tall saguaros that stood like sheriffs in the desert, arms akimbo, and the ocotillo like a spikey whip that burst into bloom in spring. In this mesmerizing landscape, I found new friends. The people I gravitated to were Renaissance men and women: writing poems, playing the cello, walking the beautiful poodle, and finding time to drink coffee and talk writing. That was the sort of richly textured life I had always wanted, and to be surrounded by people who were living it made me feel like I had found home. Sam and I bought a house, painted the trim blue, and sanded the pine floors. I imagined living there a long time.

But Tucson is filled with people who came for a degree at the University of Arizona and stayed. It's overfull with talented writers toting their MFAs. I couldn't believe my luck when I was offered a teaching position at Bard College. That meant moving back East closer to my parents as they aged, closer to old friends and the place I originally called home. But it also meant leaving Sam, the house we had bought, and the life we had created together. Sam was sailing into a new life as well. So after a few fights and lots of tears—Sam got the dog, I got the cats—I left. Still, the tug of Tucson brought me back regularly. And how perfect, it seemed now in the winter of 2012, that a place I loved, where I had friends, was also a major birding destination.

While I lived in Tucson the birds had caught my eye—it was hard not to enjoy the Greater Roadrunners that streaked through the ocotillo, the Cactus Wrens atop a barrel cactus that serenaded every hike, or the cascading song of the Canyon Wren that joined me in remote canyons. It was there that I bought my first pair of binoculars to spy a Vermilion Flycatcher and to adore the Gambel's Quail that poke about in gravelly backyards. The desert made casual birding easy, but beyond admiring these obvious birds, I hadn't yet been bitten.

I knew all along I wanted to be familiar with the birds, but I had been a hesitant, even shy, learner. In some ways, my slow path reminded me of stories of people who are friends in high school, go off and live their lives, maybe even marry, then refind each other in mid-age and fall in love. They nod and smile and claim they loved each other all along. But what ingredients transformed that love between friends into romantic love, the love that gets you up to look for birds long before dawn?

I wonder about my timing, why, from the time I bought those first binoculars in that desert city, I waited twelve years to fall for the birds. I had my reasons of course, but I was also thinking of Frank Chapman's conversion to the birds, of St. Augustine's conversion to God. Both of these moments had a witness, another person present. No one takes a leap into another life alone. Maybe I had been waiting for Peter, knowing that it would be hard to commit to the birds without Peter at my side. But maybe it was now time for me to realize that I had my own relationship with the birds; I didn't always need him at my side.

At the same time that I gave over to the birds so too did my Tucson-based friend Deb. When we met fifteen years earlier, Deb didn't hide the dinks and scratches on her strong arms and legs. "Bushwhacking into a climb," she explained, the first time I asked. That was often how she spent her weekends with her partner Larry. I envied those battle wounds and knew then that we would be friends. In the four years I lived in Tucson, we rock climbed on Mt. Lemmon, mountain biked through the desert, hiked rocky canyons. Now, she too had picked up a pair of binoculars and had gone about learning the birds with her usual intensity. Before my arrival she had sent me an email: "I'm not nearly as maniacal or energetic as you." I soon learned she was wrong.

On a January break from teaching, the second winter of my birding life, I arrived in Tucson. After a day or two birding on my own in the mornings and drinking coffee in the shade with friends in the afternoon, I was ready for a proper long day of birding with Deb. She picked me up around nine (late! I thought, but maybe desert birding didn't require crack of dawn starts) to head south to Florida Canyon in the Santa Rita Mountains. I knew the Santa Ritas well, in summer and in winter had hiked to the 9,452-foot summit of Mt. Wrightson. Even in the worst of the summer heat, it remained cool from the altitude and from the light filtered through the Ponderosa pines on the summit. One winter the mountain had surprised me with over a foot of snow. Now, I could return to these familiar haunts with my binoculars.

The Santa Ritas are legendary for birding. In *Wild America* Roger Tory Peterson wrote of an evening in Madera Canyon when he hears a sound like a "muffled puppy," which turned out to be a Mexican Spotted Owl. Then he heard or saw a Flammulated Owl, a Spotted Screech Owl, two Elf Owls, and a Great Horned. Five owl species in one place, on one extraordinary night. Beyond the abundance of wondrous owls, Madera Canyon is famous for a range of fabulous birds, including the aptly named Elegant Trogon.

On the drive down I-19 toward Florida Canyon, I called Peter. A January trip to Arizona didn't interest him, as he claimed that to see the birds he wanted to see, he had to go in August. When I lived in Arizona I tried to leave in the hot months of July and August when the heat was so intense I had to walk the dog at four in the morning; later in the day the asphalt burned her paws. Whatever birds are foolish enough to appear in late summer I would never see.

Peter was at work, balancing on a ladder, rewiring a kitchen. I could hear a drill running in the background.

"Rufous-crowned Sparrow," I said, when he asked what was the best bird of the trip so far. "And we're off to find a Rufous-capped Warbler. Everything is rufous down here." The Warbler had been reported in Florida Canyon in the past week; Deb had seen it three days earlier.

Gleeful telling him about these birds, I also felt this was some sort of strange betrayal, as if I were cheating on him by seeing birds he wanted to see.

"Maybe I should have come with you," he said quietly. I knew he was regretting not seeing these desert, wintering birds; he was not missing me. We were in a midwinter slump, brought on by shorter days and fewer birds. As if insulating ourselves from the cold, we had zipped up our down parkas and turned inward. We no longer called in Saw-whet Owls at dusk but rather sat in front of the TV while eating dinner at Peter's house. "People really watch this stuff?" I asked as we took in another rerun of *The Office*. Initially, the TV watching, with our thighs resting side by side on the couch, felt sweetly intimate. Above all, it was something new for me in a relationship. When Sam and I lived together, Sam wrote in the evenings while I graded papers. We didn't even have a TV. And though TV watching with Peter felt comfortable, I saw that it was a lot like only ever wearing sweats around the house—eventually you feel frumpy. And then what follows is that your love becomes frumpy as well.

I experienced that dullness as steadiness. The woman who had packed up and left Peter's house without a word, leaving the pumpkin pie behind—that is who I had always been, the one who left. So I was proud of myself that we were sticking it out through this winter, even if it felt monotonous. I always claimed boredom is good; it's when the imagination takes flight.

And if the imagination doesn't kick in then a plane ticket can help. Just two days into my trip and the Arizona sun had evaporated the gray of winter that had been sloshing inside of me. This was the break I needed, from snow and cold, and, yes, from Peter. Hanging out with Deb made me giddy, and the sun inspired me to laugh.

After a half hour, Deb exited I-19 and turned onto the road that led toward the Santa Ritas. We had the two-lane road to ourselves, the desert stretching out on both sides, the mountains rising in front of us. A bird perched on the electrical wires, dangling its red and gray tail earthward. "Hang on a second, that is one very cool-looking bird," I said. Deb stopped, backed up, granting me a few minutes to admire this Pyrrhuloxia, a common bird of the desert that looks a lot like our Northern Cardinal, but with subtler colors and a chunky orange bill.

"Just a Pyro," Deb sang as we whipped down the road. Pyro! That I could pronounce while Pyrrhuloxia (Pyrrhula—flame colored; loxia—crooked bill) was a syllable too many.

"*Just* a Pyro? Stop."

Then I noted another fascinating-looking bird clinging to the wires, the Phainopepla with its high black crest and a glossy blue-black coat over a long, slender body. It made sense that Phainopepla is Greek for shining robe as the bird had that sort of distinction.

"Just a Phanney," Deb said, every time I pointed one out.

"*Just* a Phanney? Stop."

Surely I had seen these common birds before. I simply had never paid attention to them. And I wondered aloud what common birds of my woods and marshes would make Deb linger with excitement ("Pileated," Deb said emphatically; "Wood Thrush, Ovenbird."). I suppose one way of thinking of this is one person's trash is another's treasure (and though birders do discuss "trash birds," I don't like thinking of any bird as trash). I imagined stopping to admire every Robin or Blue Jay, and though they deserved that admiration, I also realized I would never get to the trailhead.

After a half dozen stops for Phanneys and Pyros, we slowed for the final mile on a lumpy dirt road, kicking up dust along the way. In a small opening, we joined several other parked cars. The Arizona birding community had mobilized to see this special warbler, with a rufous crown and cheeks. It's a bird that lives mostly in Mexico, making guest appearances in Arizona or Texas.

Deb pulled out a tiny pair of binoculars, Leica 8 X 20 trinovids.

"You know, your bins look like a toy," I said. I knew that my good binoculars had made my experience birding that much easier, that much brighter. But in criticizing Deb's optics I wasn't being a binocular snob; I truly couldn't understand how she could find anything with such small bins.

"I love these," she said.

I couldn't deny it, Deb was nimble with them, pouncing on the birds, first a sturdy Spotted Towhee, with its black head and red eye. It had red flanks and sported a spotted black cape. Then she pulled a Ruby-crowned Kinglet out of a bush. The nervous-seeming, tiny bird raised its red crest, thrummed one of its cascading songs. The Ruby-crowned Kinglet is a snowbird, spending its winters in all of the southern states.

Deb and I had always shared a physical restlessness. That day it kicked us up the dusty trail, following a dry streambed. Watching Deb move through the brush motivated me; she was elastic as she jumbled from one boulder to the next. Though I appreciated my newfound ability to sit still and watch a bird, this was still the sort of birding I loved most—active, moving, pursuing.

Two parties passed us on the way down shaking their heads. No luck with the bird.

We settled on a rock while Deb read aloud from an entry posted on the Arizona Listserv describing the last spot where the bird had been seen.

We had to keep our eyes open for the bird just past a small dam. Reading the directions felt a bit like "birding by numbers," and yet it was true that birds, once they found a good spot with water and food, often stuck to that spot.

While Deb went over the directions, I looked at the image of the bird in her guidebook. Her *Sibley's* was cracked, edges rounded.

"What do you do, chew on this thing?"

"Wait till you meet the empid flycatchers."

The *Empidonax* flycatchers are notoriously difficult to identify. Sibley cautions that an ID should only be made with several characteristics, most importantly voice.

I sensed Deb was going to keep me on a diet of skulky, hard-to-find birds like this elusive warbler. Quietly, I was thrilled, in part because I had worried that without Peter at my side I would not see good birds. I had come to feel that birds and Peter were one. With a bit of distance came perspective; he always led, and I consistently followed. As grateful as I was for all of his guidance, years spent in the feminist trenches made me think this was not OK.

I had always treasured my independence, and feeling reliant, needy in any way made me itch. But it was really that in following Peter I wasn't developing the confidence I now saw in Deb—the boldness to strike out and find a bird and spend time chewing through the guidebook to figure it out on my own. I was learning all the time with Peter, but I was learning from him, not with him.

Deb was guiding me through her bird world, but it felt different, perhaps because we each brought something to the quest. From time to time she hesitated in an identification, and we pulled out the guidebook to puzzle it out together. I was relishing being with a friend with comparable skill and experience.

Together Deb and I combed the bushes, dense by the trailside. Soon, we abandoned the trail, rock hopping along a narrow streambed, scratching our way back and forth for over an hour. From time to time we stopped to listen intently, hands cupping our ears to amplify any possible *chip* or *tik*. A sweaty discouragement coated my arms and neck.

This stick-to-it would have been nothing to my imaginary birding partner and literary companion, Florence Merriam Bailey, who was never far from my thoughts. She, too, had birded this area of Arizona and would

have been tucked under a bush, ready to wait a few more hours. But neither Deb nor I had such patience, and by midafternoon agreed that the
bird would not likely sing at that hour. We gave up our hunt. I hardly felt
disappointed. The warbler would have been great—ice cream after a good
meal—but I was elated from all of the new-to-me birds and seeing this
land through my binoculars, Deb at my side.

On the descent a gray blob of a bird, with a tiny bill, yellow-olive breast,
and nothing else to make it identifiable, landed in a bush in front of us.

"A Hammond's Flycatcher," Deb said. An *empid*! "See how the tail
flicks up, not down?"

"Right," I said, so happy to meet this empid flycatcher. "Now I get why
your guidebook is a mess."

Midafternoon two days later, Deb and I were headed east and south to
Bisbee. Bisbee is a former copper mining town with a latter-day hippie,
new age twist. It drapes over a canyon, with houses perched on the side
of the canyon walls. Larry, Deb's partner, lived in Bisbee, and we would
spend a night or two with him there.

Our first morning out of Bisbee, we birded Whitewater, an open area
with water impoundments that held a delicious array of ducks. The big
draw were the Sandhill Cranes landing by the hundreds, thousands. We
joined flocks of people to admire the elegant, tall gray birds with their red
crowns milling about on the flats; the sound that emerged was like a cat
with a wooden purr, a rattle, a constant chatter of contentment. The dizzying come-and-go made me fear the birds might collide as they flailed toward the ground, all gangly legs and awkward wings, while nearby others
took to the air to circle over, check out the view, and survey the undulating
ocean of cranes, only to come to ground nearby.

After our morning in the intense desert sun, I was fantasizing about a
nap at Larry's house, just thirty minutes down the road.

"Want to go look for a Black-chinned Sparrow?" asked Deb, the
woman who had claimed to be not as energetic as me.

My heart fluttered.

"We could walk up the canyon." Her eyes trained on the road. "I saw
one there a few weeks ago."

"A Black-chinned Sparrow?" I had been seeing Black-throated Sparrows throughout the trip. (Chin? Throat? How irritating that someone

had not been more original with these names.) This cousin was a scarcer bird. I was now fully awake; that nap could wait. A special *sparrow*. Now we were really birding.

Deb navigated the narrow, winding road and soon pulled over to park, so close to a stone wall I thought I heard the paint peel off. Once on foot we passed a range of makeshift homes, perhaps old miners' cabins. A copper-colored dog joined us, cheerfully loping ahead, then returning to us as if we had always belonged to each other. Stone walls flanked the trail where grasses grew, shaded by a mesquite tree or two. A few Vesper Sparrows darted from the bushes. Further up the canyon it became fantastically quiet. Not a bird in sight. The sense of calm, the shifting, dropping light, and the cooler temperatures after a hot January day, all of this soothed me. But with no birds about, we headed back downhill to the point we had last seen a bird. We stood. We *pished*. We gazed into desert brush, at small crumbling cliffs.

"There it is," Deb said.

The elegant compact gray bird with a faint black splotch on its chin appeared like a vision in a scrubby bush, announcing itself to the world: Black-chinned Sparrow. It sat, as if greeting me, enjoying being adored for a few moments. I was excited, not just because it was an "uncommon" bird, but because it was a sparrow, those little, uninteresting, overlooked birds that were Peter's specialty.

"Thank you," I said as we retraced our steps toward the car, elated. An adrenaline buzz coursed through my body. "You are so good," I congratulated over and again.

I felt a wave of affection for my friend. If I had been on my own, I never would have known to look up this lonely canyon for the sparrow. She was a great guide to the desert. But beyond great bird locations, four eyes are always better than two, and her eyes were sharp. If I wanted independence from Peter, that didn't mean I had to bird solo.

On this trip, as Deb and I drove from one birding location to the next, at one point stopping in a charming coffee shop in Patagonia to buy tea and one of the best BLTs I've ever eaten, we roamed through the subjects of life, love, and work. The quest for birds gave us time to know each other better. When I lived in Tucson I had been living with Sam, had been a graduate student. Now I taught full time and had a boyfriend. Deb teased me about Peter, calling him Petie as if they were already friends. We always returned to birds and the places we would see them—Peter and I were off to Alaska in June, and Deb dreamed of South Africa. Above all, I loved hearing Deb laugh when we found a good bird.

Once we were back in Bisbee and had cell phone reception, I called Peter. I felt a tingle of excitement as I pressed the phone to my ear, like the student who has her first short story published, then calls to tell the teacher who most inspired her work.

"What did Petie think of the sparrow?" Deb asked.

"He said 'good bird.'"

"That's it?"

"I know." But I didn't really blame him. By now he had to be kicking himself for not coming along on this trip.

Deb and I wandered the streets of Bisbee wanting something celebratory to mark this perfect day, the special sparrow. I loitered in front of a jewelry store, admiring necklaces and bracelets that had that southwestern feel, turquoise and gold marrying in original pieces. I'm not a person who wears jewelry beyond a ring or at times a necklace, but still, the store drew me in. I slipped a wide gold band with a desert turquoise embedded in it onto my ring finger and surprised Deb and the owner by saying I wanted it.

"I married the Black-chinned," I joked as we walked out of the store. Was the joke a cover? A bit, I sensed. My commitment to my writing, my teaching, the birds was complete. Maybe now I wanted to commit to a person, to Peter. Or maybe the joke was that I had married myself. If so that was kind of wonderful and sort of appalling. Wonderful because we know that love begins with self-love. Appalling because it felt far too self-involved. Maybe the ring didn't mean anything at all except that I loved the feel of it on my finger, the color of the turquoise. And I knew that when I wore it I would think of that little sparrow, feel the desert air, remember Deb's laughter as we chased those desert birds.

Back in Tucson, I felt enormously satisfied by my trip: birds, friends, good food, perfect weather. All of it had changed the bad attitude I had arrived with into renewable energy.

On my last day in town, I met Sam, who had stayed in Tucson when I left. It took me a second to recognize him, seated at a table with a cup of coffee. In the years I had become a birder he had transitioned, now living as a man. I did not equate these two life changes. I was thinking, though, about creating lives, composing lives one bit at a time: the years before Sam transitioned and now his living fully as himself; the years before birds and now the years with birds. Sam was Sam before his transition, and the birds were with me before I gave over to them. To give over to our true selves made us both bigger, more inside of our own lives. Like the birds, we had become who we were, who we had always been. Understood this way, I realized how birder had become such a big part of my identity, how when I introduced myself my identities of teacher and writer trailed after birder. It wasn't just an idle label; it was who I was, down to my toes.

Over our late morning coffee, I admired Sam's courage, and also worried for him, a small man, a poet, in a not gentle world. Seeing him, though, his thick-rimmed glasses and five-o'clock shadow, I was able to feel his solidness. I was so grateful that Sam was able to be Sam down to his toes. He bragged about being able to do pull-ups with ease. I saw, talking to Sam, that we were both happier than when we had lived together in that airy house in Tucson.

Sam told me about his partner, and I told him about Peter. On the surface Sam and Peter seemed so different: my Jewish intellectual poet-artist, and my electrician-photographer-birder. Had I really changed that much? And yet Sam's focus on his art—photographs, poems, music—matched Peter's focus on the birds. Both had undergone a journey to find themselves, and that journey had involved painful choices and loss, Sam of his family, Peter of his son. They lived with these wounds. But with intelligence, imagination, and determination they both had transformed that pain into vision, into the lives they wanted to live.

As Sam and I hugged goodbye—he was off to a trans poetry gathering—my phone rang, Deb calling to ask if I wanted one last shot at the Rufous-capped Warbler.

"I'd be up for it," she said. This would be her third trip to see the bird.

So that last day in Arizona we returned to Florida, arriving at the lonely canyon just after noon.

"This is not a proper birding outing," I told Deb.

"Yeah, but the birds don't know that," Deb said. She had no idea how grateful I was to learn that Peter's rules of birding were not rules at all. Good birds did not always demand you get up at five in the morning. It's not that I don't like getting up at five in the morning, because I do: the alarm ringing in the dark signals a special day awaits, gives me a sense I have a mission. But I didn't like the regimented sense that without the early start, I might as well not go at all.

The sun had crested the mountain, touching down into Florida Canyon. The sky stretched a shade of blue found only out West, a blue that reflects empty space. I followed Deb, enjoying her swinging gait as we passed a wash where sycamore trees grew strong. Through a gate and up the rocky trail we trod. All of this was familiar from our earlier bid to see the bird.

A trio of birders in sweaty T-shirts and with dusty legs emerged from the canyon. They all shook their heads glumly. "No warbler."

This didn't dull our enthusiasm. Somehow I sensed that Deb and I were a golden duo. After all, together we had seen the Black-chinned Sparrow and dozens of other wonderful birds in my short week in Arizona. I spun my ring on my finger.

We crossed over a dam, water trickling through shockingly green algae and watercress that laced the edge of the spill. We tromped the edges of a narrow stream, thick with brush, greedily thriving on the precious water. Above us soared White-throated Swifts chattering in the still air.

A photographer, cameras dangling from weary shoulders, passed us. "It's not here."

Nothing could dull my optimism. While the photographer headed downstream, Deb and I climbed higher.

Deb squatted on a rock, peering with her tiny bins into the thick brush. "Got it," she said. Her hat covered her face from the sun, but I could see the determined smile. She looked like she might roll into a somersault or spring into the air, but she stayed compact, her binoculars to her eyes, unwavering. I followed where her gaze traveled and spied the bird. The dull olive-colored warbler with a striking rufous cap flitted from one bush to the next, allowing only the briefest looks.

I sat back on a boulder as if I'd just performed an athletic feat.

"Good bird," I exhaled as Deb and I burst out laughing.

GUIDED

Gambell, Alaska

"Siberian Chiffchaff." The voice crackled over the walkie-talkie.

The room exploded, everyone grabbing coats, slapping on boots, strapping on binoculars. Chairs toppled over as people rushed toward the door. Twelve people, all in various stages of after dinner exhaustion, sprinted out into the late-night sun of Gambell toward the boneyards.

This group of Chiffchaff chasers was part of a guided trip to Gambell, Alaska. Sometime late fall Peter had said Alaska, and I said yes. Then he said, "Let's do a tour, and go to Gambell," and I hesitated. I didn't like the idea of guided anything. If my trip with Deb had given me a taste of independence, a guided trip was heading in the wrong direction.

Above all, I was hesitant about Gambell because in the birding world Gambell is Compostela, Finisterre—the end of the world. I hadn't walked far enough, prayed hard enough, did not have enough faith. Yet once we determined this was where we were going, had paid for our trip and started gathering gear, worrying about cold in June, Gambell became my own holy destination. Laced through my excitement at traveling to a place so different, so remote, was a lingering concern that I was going to be flummoxed by the rare birds that catapult themselves from Siberia to this remote town on the edge of St. Lawrence Island. The voice crackling over the walkie-talkie on our second night in Gambell confirmed my worry: *What was a Siberian Chiffchaff?*

That didn't matter. Caught in the group, I trotted down the gravel road past a few square wooden houses as I zipped up my parka. The temperature hovered in the upper thirties, taking me back to winter gloves and hats. Turning off the road, we pointed in a single-file line along the edge of the rocky, crater-like boneyards that form a rough-edged border to the town. In the near distance stood Paul, binoculars to his eyes. A bird guide who had settled into Gambell for a few prime weeks, Paul had the bird staked out.

Before arriving in Gambell I had read and spoken about the boneyards but couldn't picture the landscape, the pits of gray gravel and bone, chipped, cracked, and worn smooth white or gray. These pits contain the remains from the Alaskan Native subsistence hunt: walrus, seal, sea lion, and whale. After centuries of hunting, the bones have mounded up, forming junkyards of bones, layers and years deep. The Alaskan Natives dig through, finding pieces of ivory or fossilized bone to carve, to sell to buyers from shops in faraway Anchorage. The Alaskan Natives of Gambell, known as the St. Lawrence Yup'ik, also came to where we stayed, the Sivuqaq Inn (Sivuqaq is the Yup'ik name for St. Lawrence Island), in the evenings, quietly slipping the carved bits of ivory out of the pockets of their heavy down jackets to display on the long foldout tables where we ate our meals. The carved pieces ranged from delicate birds to the solid sea lion jaw Peter bought me. One evening two smiling men arrived with a smooth, white bone, two feet long and gently curved with a blunt rounded end. A fossilized walrus penis. I had to have it. Everyone laughed as the men handed over the bone called an *oosik*.

While the Alaskan Natives made a little money carving what they found in the boneyards, we took advantage of the boneyards in another way. Into this rocky, uninviting landscape birds drop, perhaps blown off course as they flew the skies between Gambell and Siberia. Where a bird might normally take refuge in a bush or a tree, all a bird had in treeless Gambell were these pits. The birds tucked in behind a rock, blended in with the gray, white, brown broken-bone mixture, then sat still. We would walk between the pits, bones crunching underfoot, and peer in, hoping to see movement, life. Searching the boneyards was like a daily Easter egg hunt: you never knew what you might find. Now here was this Chiffchaff—maybe.

I watched as the expert birders flushed the bird. I could see a small, nondescript speck, which one guide explained was a warbler, as it flung itself from one pit to the next. Every time the bird moved, everyone with a camera frantically snapped pictures; a photo would be essential in confirming this ID. One guide kept us tourist birders at a distance, pointing when the bird made a brief appearance. And brief it was. I would see a dull brown something emerge, a surge of humans lunge toward that spot, then poof, gone. The best thing about the bird was its name. Chiffchaff. Chiffchaff, I silently chanted to myself as I breathed in the cold air of the north.

"I don't want to lose our focus," the tourist tamer said, "but over there is a Yellow Wagtail."

The uncooperative Chiffchaff lost my affection to the Wagtail, confident and yellow, that perched erect on a mound of bones. It was equally exotic to me, and to my pleasure, I had the bird to myself. For a moment I forgot the group of other birders as I admired this bird whose close relative, the White Wagtail, Peter and I had met just six months earlier on the banks of the Seine in Paris.

Soon the Chiffchaff vanished into the thin air of the north leaving behind the big question of was it really a Siberian Chiffchaff. No one was certain. We all draggled back to the inn, to the open gathering room with long plastic tables and foldout chairs. Down a hall were our Spartan bedrooms with twin beds and little else. Further down the hall was a shared bathroom. It all reminded me of my college dorm.

The top birders clustered around a computer screen where unimpressive photos of the bird led one to say Willow Warbler then another Siberian Chiffchaff. Both are dumpy warblers, greenish on the upper parts and whitish underneath. Most often these subtle birds are identified by voice, habitat, or geography—three things we couldn't use here. What we had were photos, some of which Peter had managed to snap, even from a distance. The group moved from one computer screen to the next comparing photos. "Don't delete any of these photos," our guide admonished Peter. "Even if they are crap."

The evening drew long with discussions of the length of the primaries, the flight feathers. In the Willow Warbler, there is a slightly longer primary projection. We went to bed not knowing. The excitement of the find, followed by the unidentifiable bird offered exactly what I expected from Gambell.

In my pre-trip reading, I sensed that Gambell was an Alaska few get to see, more rugged, more remote, more filled with adventure. Fueling my fantasies was the chapter on Gambell in Kenn Kaufman's 1973 memoir, *Kingbird Highway*. It was a book Peter had read three times, following Kaufman as he traveled through the country on a footloose and thrilling "Big Year."

A Big Year is yet another birder's game, where you take a year and see as many species as you can within a particular area—whether county, state, continent, or even world. Tales of Big Years are the makings of best-selling books; people become fascinated by reading about (or watching a movie about) people creating a giant carbon footprint to see special birds. Big Years are the opposite of my Florence Merriam Bailey sitting in a field waiting for the Bobolinks to sing.

It is also the opposite of what Kaufman did when he took on North America in 1973. He was not hopping on planes or hiring helicopters to take him to the summit of mountains to tick off one lone bird. He was carless and near penniless; he mostly relied on hitchhiking to get around. As he hitchhiked his way across the country, there were a lot of miserable nights complete with snow and rain. He remained tenacious. And nuts, in a determined nineteen-year-old way. In the end, he saw 671 species.

Early on, the legendary Ted Parker gave him advice: "You can hit 650 easily if you get to Alaska." Since Kaufman's goal was to beat Parker's record of 626, Alaska became his Compostela. The AlCan, the lone, 1,500-mile-long road leading to Alaska in 1973, was marked by a sign that read: Pavement Ends. The route was lined with other hitchhikers and traveled by RVs, unhelpfully passing him by. Kaufman, thumb out, eventually landed a ride and soon made it to Fairbanks. His heart was set on Nome, where reports floated in of a White Wagtail and a Great Knot. That's when he learned of a place called Gambell with a dizzying list of special birds: Red-throated Pipit, McKay's (now Thick-billed) Bunting,

Brambling, Rustic Bunting, and the legendary Ross's Gull. Reading this list of birds made my heart soar—some I had never heard of before and none had I ever seen. Kaufman got himself on a plane out of Fairbanks to Nome and on to Gambell. He was not disappointed.

Hitchhiking is how you travelled if you were young and didn't have money in the 1970s. That's what I did to get to the cliffs I wanted to climb, often thumbing my way from college in Colorado Springs to Eldorado Canyon outside of Boulder. It was easy traveling because a woman alone or even with a male friend never waited long. The summer after my first year of college, my boyfriend Michael and I thumbed our way from Colorado to Tuolomne Meadows, California. We got there in record time, truckers bringing us along for long hauls. They wanted us for company, to listen to their stories or in turn to keep them awake by telling our own. I remember one trucker, trim and long-legged, the inside of his cab impeccable, giving us a list of all the gifts he had bought his wife, which included a fur coat and a refrigerator. "And she's still not happy," he moaned. *Of course she isn't*, I thought. The only thing that made me happy then was climbing, so the idea of objects delivering happiness seemed absurd. It still does.

Michael and I spent every day in Tuolomne scaling the beautiful smooth granite domes. We were so wholesome looking that mothers with children picked us up. One day a French couple from Bordeaux getting around in their rented Lincoln drove us to Yosemite Valley.

Out of my youth I've constructed these fun, I-was-so-lucky stories, just like Kaufman has. It's easy to look back and laugh, because I'm still alive. But when I include other stories, like the ride with the drunk man going ninety in high winds out of Vedauwoo, Wyoming, or the many times we were off route, or took long falls, the days when the sun went down and we were far from a meal or our tent, when I include that because I was poor and wanted to be light in order to climb hard I became unhealthily thin—it seems a wonder I survived.

Every day, Michael and I navigated the sheer granite faces without a guidebook, searching for bolts like playing connect the dots with our lives. It was there on those domes that I learned how to trust my feet on razor-thin edges, and led routes far harder than I had ever imagined. My sharpest memory is of looking into a blue sky toward Michael, chest bare, red swami belt around his thin waist, long bowed legs stepping on thin holds as he danced past fear. We left Tuolomne when all of our fingertips had split and were bleeding, too sore to hang onto anything but a cold beer.

I was too young to understand the extent of Michael's daring; he led us up the hardest routes, ones with so little protection falling wasn't an option. Years later, when I learned from a friend that he had taken his own life, I sat in my small kitchen in Tivoli, and a well of sadness filled me. Such a smart, beautiful man with his curly mop of blond brown hair and crooked smile, and more fragile than I knew, more afraid than I ever imagined.

I often wondered how my parents sat back and let me go into this world of climbing. Once I was an adult and he an aging man, I said to my father, "I'm surprised you didn't worry." He looked at me aghast, his six-foot-four frame deflated with a sigh that verged on tears. "Worry? God I worried." Never once did either of my parents say: don't go, don't do that. Rather, when I graduated high school and headed to climb in Colorado, what my father gave me as a gift—to add to the pile of slings and ropes, cans of beans and bags of rice packed into our VW bug—was a copy of *War and Peace*. The idea that I would read this epic novel while camped in a parking lot after a hard day of climbing now only makes me smile. At the time my thought was: my father has no idea what I do.

"Do you understand we never could have stopped you?" my father said. Unaware of my own determination, how much I needed climbing, each route my way of communicating with the world, I also couldn't see that for them watching me climb was unbearable.

The one time my parents came to the cliffs, the Gunks, I tried to impress them by waltzing up a delicate but moderate (I did not want to risk a fall with them watching!) face climb with the lovely name of Pas de Deux. After I finished the climb, I tied off my ropes and rappelled to the ground, proud and chalk-dusted, little biceps bulging. My mother did not marvel over my feat, exclaim over my grace. She asked: Where are we going for dinner?

Were my parents alive, how relieved they might be that I had now found a passion that on the surface seemed so tame. And yet birding, I learned from Kaufman, could also be very not-tame, which is why I was so taken by his story. In his telling, birding is for the young, the adventurous, has its own unexpected dangers. The story that most gripped me was when, at the end of his Big Year, he participated in the Freeport, Texas, CBC. Kaufman's sector involved sitting out on a jetty to get the rare seabirds. The waves crashed and roiled around his promontory, licking at his boots. But! He saw that Black-legged Kittiwake. It's while he was scoping a Jaeger, trying to decide if it was a Pomarine or a Parasitic, that a large wave scooped him and his borrowed scope into the ocean. The violence of waves along a jetty is something you don't forget; that Kaufman was not instantly dashed against the rocks seems a wonder. Weighted down by his boots and clothes, Kaufman was washed up onto the jetty where barnacles sliced into his palms as he lunged for the rocks. It's a passage almost unbearable to read. He got himself out of the water, cut and bleeding, then returned to town to bandage up. Rather than sit inside and feel thankful that he was still alive, he headed back out to see what birds he had missed. I learned from reading Kaufman that a certain kind of birding was not for the faint of floppy hat; there was real adventure to be had. And I knew from two previous trips to Alaska that it is the land of (sometime dangerous) adventure.

Yet I was sure that all of our experiences were on mute as long as we were guided, and I wished we were having a more free-for-all trip like Kaufman. With a bird guide to escort us to the bird and then to identify it, a guide to lead us to the next meal or hotel room, I felt tamed. On a guided trip, I would miss my nights spent shivering in a tent, the wrong turns and unfortunate meals. Without these challenging moments, I worried there was no story.

In all of my years of climbing I had never hired a guide. I had learned to climb by making mistakes, by falling. I read Royal Robbins's *Basic Rockcraft* and trusted those who had more experience than me. This was an age before guides were common, and in any case, I hardly had the money for gas to get to the cliffs let alone a guide. Taking a youthful high road, I decided that those who did hire guides were dull, maybe not even real climbers. In both climbing and birding, using a guide felt like half a commitment to the quest. It felt, a bit, like cheating.

Still, I had agreed to this trip. I had capitulated when Peter read me the section in the guidebook about Gambell. We needed permits to walk the land that is owned by the Alaskan Native community. We had to make arrangements with the Alaskan Native Corporation to stay in what served as the hotel. There are no restaurants. The community store sold overpriced canned goods and soda. Travelers arrive with coolers filled with food. There were no cars to rent, only ATVs, which we could hire to take us to the farthest birding spots. It sounded logistically complicated; a tour made sense. Since we had sold out, we added Nome and the Pribiloffs to the itinerary as well.

On this first part of the tour in Gambell, we had the early afternoons off, to walk or, for many, to nap. One afternoon, Peter and I wandered out toward the boneyards, passing several wooden huts. Next to them stood wooden racks where meat—strips of seal and walrus—draped, in varying stages of drying, of turning from orange to black in the Arctic sun. Next to these meat racks were mounted satellite dishes, beaming television into the wooden homes of the 160 Alaskan Natives living in Gambell. Everywhere empty macaroni boxes, Coca-Cola bottles, and plastic in all stages of being plastic drifted in the wind. To say that the contrast of this junk with the subsistence life was jarring to me is an understatement. I spent hours trying to accept that we shipped Coca-Cola to this remote island.

It felt good to walk with Peter, just the two of us, after a few days of the logistics of group movement. I stopped by a post out near the water. On top of it perched a walrus's head, neatly sliced at the neck, eyes shut, whiskers stiff, glistening in the sun as if the animal might be smiling into the Bering Sea. I pulled out my pocket camera to take a photo.

"Bitches." The call emerged like a shot from inside the wooden door-frame. I turned my back, tucked my camera in my pocket.

"I read somewhere you aren't supposed to take pictures of their food, their houses," Peter said.

"Why didn't you tell me?" I felt awful.

Our guides told us not to bring alcohol as Gambell is a dry town and, once on St. Lawrence Island, not to go up the mountain outside of Gambell where they bury their dead. We were also forbidden to walk past the end of the far lagoon, which stretched from the airport and on out of town. Had I missed something?

"Hey. Hey. Hey." A young man ran toward us. He wore a heavy North Face down jacket, black Levi jeans. A cotton cap covered his head.

"Sorry," I said by way of introduction.

"That's just my uncle," the young man said, gesturing back at the wooden house. "You can take a picture," he said. The young man's voice had that lilting quality I heard with all of the Alaskan Natives of Gambell. Whatever he said, it would seem gentle.

"I didn't mean to offend," I apologized again.

"It's OK. Hey, you guys looking for birds?"

I held up my binoculars; Peter lifted his camera clutched in his gloved hand.

"Were you born here?" I countered his obvious question with one of my own.

He nodded.

"Have you ever left?"

"Once," he said. "My cousin had leukemia. He made a wish."

The stories Peter had told me of his son Forrest making a wish, only three years old, came to me. His wish: to visit the set of *Blue's Clues*, his favorite TV show. So he and his mother and father climbed into a limo and drove to New York City for one magical day.

The idea that the Make-a-Wish Foundation stretched to the far reaches of our country warmed me. But it also seemed a bit odd, this community so isolated. Whatever the cousin had wished for was not a limo ride away.

"He wanted to see a Lakers game. So we went to LA." He paused for effect while my mind tried to imagine what it was like to go from Gambell, a place with no cars, to Los Angeles, a place I think of as wall-to-wall cars.

"The Lakers won." He smiled so wide that it made me want to hug him. It also made me want to take him to the dentist.

The young man's cell phone rang, an odd sound in the high thin air. He skipped away, ear to phone, waving as he went.

Shaken by our encounter—despite the young man's words of forgiveness I still felt bad about my cultural misstep—we decided to leave the town and explore the gravel road that led to the lagoon. We passed the rectangular loaf of a mountain that brooded over the town. The gravel rolled under my feet, somewhat like walking in sand, feet flexing and calves stretched. My gait was jumbled, lurching, as if I might be drunk. Soon, we opted for the hard-packed gravel that serves as a road for ATVs motoring along the flank of the mountain.

Arctic ground squirrels, alert like prairie dogs, stood up, then vanished into holes, and tundra voles scurried for their lives. Arctic willow, the largest plant around, grew like a miniature pussy willow, a few inches high.

Western Sandpipers tiptoed through the wet spongy areas. Snow Buntings sang from atop boulders. I remembered the Snow Buntings seen back home in New York State and marveled that they might be (though probably were not) the same birds. They looked so different, these in breeding plumage, with a sharp white body draped in a striking deep black cape, while the ones at home were off-white with a wash of brown.

For the Norwegian Fridtjof Nansen, one of the great explorers of the Arctic, the song of the Snow Bunting after a dark, icebound winter brought him hope. I could see why: they were like the cheerful chickadee of the north, the bird always about and always in a good mood. The birds whistled past us, then crouched atop houses, like miniature pigeons. They seemed proud and oddly dazed by their own song, which rang out in the empty air.

Late that afternoon we walked out to join our group as we convened at Cape Chibukak, the northwest point of St. Lawrence Island, to sit on the gravel beach, watching flocks of birds surge by. The various bird formations were at once chaotic but orderly, the line of birds an express train steaming north. Forty miles away, across the Bering Sea, glistened the white mountains of the Chukotsk Peninsula of Siberia.

We sat in a scattered row in our heavy parkas, legs stretched out. It would seem that this time sitting and watching would be a good moment to get to know the others on this trip. I had vaguely hoped to find kindred spirits, people I might visit later in their homes in great bird locations like Texas or Florida. But that hopeful sense dwindled quickly.

Most traveled as couples: the quietly focused duo from the Midwest who studied the guidebook like it might be the Bible, or the latter-day hippie couple from Arizona who sweetly floated through the days. There were a few singlets, like the lawyer who wrote short stories on the side, or the energetic seventy-year-old woman from Texas who never missed a bird, or the biggest Bonnie Biglist I'd ever met. She never stopped talking about her accomplishments, yet frequently asked, "now, what's that?" to near groans from the other travelers. But we didn't talk to each other sitting on shore. And I was glad of that as the stream of birds so mesmerized me that I wanted all of my attention focused on the erratic stops, the elegance and anarchy of movement, the rush and determination of these birds forging north.

Thousands of birds trace the contours of land, rounding this last bit of St. Lawrence Island before arriving at their farthest north to nest.

It was like watching Ocean TV. Arctic Terns, Black-legged Kittiwakes, and Glaucous Gulls galore moved in time with the swells of the ocean, roller-coasting toward the horizon. The Horned or Tufted Puffins with their chunky orange bills flew in formations that undulated, collapsed, and re-formed like a messily choreographed dance. Thick-billed Murres, their black-and-white bodies distinct, often led these water-bound charges. The Least Auklets, flashes of black and white, flew in clusters, low to the water. They would land with a *kerplash*, rest up a bit, then move on like racers fresh out of the starting block.

Midwatch, a pair of Spectacled Eider touched down in front of us. I jumped to my feet to get a better view of their clunky bills and crazy eyes, which seemed to be peering up from the bottom of a snow cave. We all scrambled, the gravel like ball bearings beneath our feet, to get a clearer view. In the shuffle, Bonnie Biglist slipped. I leaned down and reached out a hand. When I next looked out, the birds had flown off. Until that moment, I had never once in my life felt regret at a kind gesture.

Seeing all of these birds was like going to a big party where you meet a lot of rich and famous people who don't care a bit for you and whom you'll never see again. It was exciting and disorienting. The blur of birds was like a dream, one that is animated with vivid colors, but in the dream I am not able to move or touch anything and I'm struggling to focus.

What I had learned and loved from the start of birding, whether at home or in Wyoming or Florida, is that birds let me know a place more deeply. In Gambell that sense of intimacy vanished. We, like the migratory birds or the vagrants, were just passing through, so fast I would not get to know this island or the town we stayed in. Though I didn't like how rough that felt—the distance it created between me and the place—it also felt right: this land was the province of the Alaskan Natives living their subsistence lives.

Usually when I travel, I play a game, imagining I live in the place I am visiting. I construct a routine of where I take a walk, where I shop for vegetables, what I do on the weekends. With Gambell, it's not only that I couldn't imagine living there, I did not even try.

Mornings were mercifully quiet in Gambell. Even the dogs slept, draped across wooden doorsteps in front of often brightly painted wooden homes. Through the night, the noise of ATVs ratcheted up, reaching a peak around

3:00 a.m. Where was everyone going? I wondered. Of course I also wonder this when I drive the Garden State Parkway through New Jersey.

Later in the day gunshots added to the roar of the ATVs. This frenetic, sometimes cheerful energy was not what I had expected to find in Gambell. I had hoped to find a place that had slipped the yoke of time, one not dictated by the clock but by a sun that didn't set except for a vague gray period around 2:00 a.m. I would tune into an inner clock, one that was about listening, to the birds, to myself. Was I hungry? Was I tired? Was that a murrelet? Did I love Peter? Not the Peter on this trip, often impatient with our group and with me. "Time's up," he said as I peered through the scope trying to identify a bird. *Time's up.* Maybe it was.

The birds and the pace of the trip did not allow me to mull that over, as we trekked out across the rocky landscape, toward the far end of the cliffs. Our goal was a Dovekie, the Little Auk of the north. Two days earlier we had looked for the bird with no luck. Now our guide had found one, staying up all night, scouring the cliff. *Crazy*, I thought.

The Dovekie is a black-and-white football-shaped bird with a stubby bill. When it swims it has no apparent neck. In its black-and-white cloak, to me it resembled all of the other black-and-white birds that zigzagged about the cliff: Thick-billed and Common Murres, Horned Puffin, and Least, Crested, and Parakeet Auklets. I focused on the goofy crest of the Crested Auklet, and the orange stubby bill and nice eye line of the Parakeet. The mountain squawked, chirped, and barked with bird song as the birds made their nests in cracks or fissures in the rock. I knew I was witnessing something fantastic and primitive, something I would never experience again. I let the awkward symphony of avian love wash over me as I scanned the cliff.

"Will I know when I see a Dovekie?" My eyes ached from peering through the scope.

"You'll know. It's unmistakable." To our guide it surely was.

While we admired birds at one end of the cliffs, at the other end boys from town shot them. Later one told us, "Auklet tastes good."

Most on this tour were active listers, all wanting that Dovekie, and several in the group reached the thrilling number of 700 with a White-tailed Eagle that rowed across the sky, above the far lagoon, wing beats heavy as if waterlogged, until it became small, smaller, a speck, then disappeared like a majestic magic trick into the slate gray sky.

It was reading Kaufman's *Kingbird Highway* that first got me think-ing about listing and the ways that people can care about that list. While visiting Gambell, Kaufman's list grew, with a few gems, like the shocking-white Ivory Gull that appeared with its big dark eyes that then melted into the clouds. Though Kaufman was list-bound, he remained clear that what was important were the birds, not the list. The list was the motiva-tor to get out, to travel. And this traveling was all about the journey, not the destination. On his trip to Gambell, Kaufman was so immersed in the journey that he forgot his list. When it finally came back to him, he real-ized that he had passed 600.

Too often, listing is a box checked off, then forgotten like the trophy gathering dust in the spare room. But in Kaufman's interpretation of list-ing, each bird has a story. In that story our lives are richer. I did feel that with each bird my life gained depth and texture. There in Gambell, however, I sensed that the story was weaker, less interesting, because the bird was handed to me. A collective *oooh*—relief, gratification, awe—rose from the group when we finally trained our scopes on the Dovekie.

Our final night in Gambell, I was out, propped on the back of an ATV, returning from seeing a Red-necked Stint at the far end of town. Peter had stayed to photograph the bird, capturing this little sandpiper from Eurasia with its distinct red neck and bright rufous wings. Tired and chilled, I opted for an ATV ride back into town. I straddled the padded seat, the wind in my face. It was fun, in fact, to finally be riding one of the ATVs.

At one point I had seen a teenager driving along with one sibling cra-dled in her lap, another clinging on from behind. When she hit a dip, the child behind bounced off, landing hard in the gravel. Without hesitation, she climbed back on, and they sped over the next hill. I held on tight to the driver as he raced across the rocky terrain as I doubted I'd rebound so cheerfully.

The driver pulled over near the airstrip. He swung his leg over the seat and stood, eyes on the horizon. He was a thick-set middle-aged man, his face calm, handsome, I thought, with high cheekbones and dark eyes.

"If we watch here, we'll see whales," he said with a wink. Was he showing me a special spot, or did he take all of the girls bundled in their sexy snowsuits here to look for humpbacks? The Bering Sea, emerald cold, spread before us. I shifted my weight, gravel crunching under my boots, to keep warm. He stood still like a Buddha, as if channeling silence. I tried to follow his lead, let my hips sink, my shoulders relax.

"What do you make of all the birders that come here?" I asked. We had to seem a bit ridiculous flying to their village to marvel over birds they wanted to eat.

"Oh, you've been coming here for so long, since the seventies, we're used to it," he said. Familiarity does numb one to what is odd.

"You ever leave here?" That was the question I put to every person I spoke to. I wanted to understand a bit more what kept the people of Gambell tied to this beautiful, austere bit of land. Once a person has tasted fresh blueberries or crunched a just-picked apple, I doubted she could return to a place with no fresh fruit or fresh vegetables. Home, of course, is more than food, but that was my starting point. I laughed at myself imagining an Alaskan Native woman saying in response: A place with no seal? No whale meat? How is that possible?

He stared out to sea giving a solemn nod. "But I came back. There's something special about the subsistence hunt," he said. "It brings us together; it's our spirit."

My meditative ATV driver didn't feel the need to explain any more than that. Why should he? I didn't feel any need to explain why I had traveled thousands of miles to see birds.

I nodded. The cold allowed me to believe I did get what he meant, his people united in their courage to live such spare lives. They shared the same air, same breath. There are many ways that a person can belong in this world: to a person, a family, a community, a town. How grateful I was to have seen, to have met someone who could show me—and articulate in so few words—the beauty of belonging to a community. I thought of those flocking birds moving along the coast. Surely they, too, knew the beauty of the whole, how it's safer, but also more fun.

How I yearned for that sense of shared spirit. It would never leave me, this sense of wanting connection—to a person, a place, or a community. Something unformed, empty took hold of me as I stood on the rocky shore of Gambell. The place accentuated everything, including all emotions. Here, seemingly at the edge of the world, my sense of aloneness felt like the open jaws of that whale we were looking for: I could be swallowed by it. I took a half step closer to my guide and resisted the urge to put my arm around him.

I'd been *the best girl*, I'd been *sweets*, and in those relationships I understood that that was good, better than good: the best that you can expect once your mother has died and you know that you will never again
be the center of someone's world. There in Gambell I understood that
being with Peter did not remove the cloak of aloneness that draped over
my spirit, and neither did this odd family of birders. What others did, so
seemingly naturally, coming together in community, I did not seem to be
able to do. Was it unusual that I felt less alone with *birds*? Maybe Olive
Thorne Miller was right, they were my brothers.

After waiting for a while longer for a whale to breach, we gave up.
I climbed back onto the ATV, my legs stretching wide to straddle the padded seat. I wrapped my arm around the driver's thick waist and rested my
cheek against this nameless man's back, holding tighter than I needed to
as we roared into town.

CHIUIT

Nome, Alaska

Summer of 1983, just out of college, I worked for a hot-air balloon company. I topped off the tanks with propane, vacuumed the van clean, then in the late afternoon in an expansive field, held the line tight as the enormous balloon filled, bellowing to life to rise like a balletic elephant. I waved goodbye, checked the direction of the wind (lick a finger, hold it high), then scanned the map for how I might track the balloon's trajectory through the sky. They floated, insouciant, above trees and houses, cows and ditches. In a balloon, the constraints of the world—roads or stop signs, dense woods or miasmic marshes—vanish.

Those balloons moved as freely as nothing I had ever experienced before. Every evening as they flew in the dusk-settled winds, I thrilled at the brilliant show of the colorful balls suspended in the sky. "My balloon went high, caught a draft," I would brag, as if the balloon were an athlete outcompeting the others for distance, or time in the air. It was also a way to explain why I was the last van back to the house where the crew lived and where we fed our clients a late evening meal.

Then one day I got to ride in the balloon. The pilot, a tanned, trim British fellow, all confidence, kept telling me to *get down*. My instinct was to lean onto the padded edge of the wicker basket and hoist up and over, as if pulled out, into the air. I was drawn out of that basket by the void—the French know this well enough they call it *tirer par la vide*; I wanted to tumble into emptiness, take wing, *fly*. Floating in a balloon is as close as I have ever been to experiencing flight as birds do, even though my flight was limited by a big bag of hot air.

What I remember most about my ballooning experience is this: I did not feel or hear the wind because the balloon is the wind. In a balloon, except for those moments of the burner blasting, silence enveloped me. Unlike moving in a car or plane, where you sense the ground passing beneath, in a balloon you don't feel like you are moving, rather, the land comes at you.

I don't know if that is how birds experience the world. What I do know is that when you look at, listen to, delight in, are baffled by, and marvel over birds, the world comes at you, bigger, more complex, more wondrous. If you are in a place where the birds are bigger, more complex, and more wondrous that sense is magnified, becomes, almost, overwhelming. That was the case in Nome, Alaska: the world and the birds came at me.

On our guided trip, Gambell had been the place I most wanted to experience, but it was outside of Nome where my most hoped-for bird of the trip lived. The Bristle-thighed Curlew.

Our tour group landed in Nome on a windless afternoon. After the shuttle dropped us at the hotel, which was disappointingly just like any hotel in the Lower 48, I headed out on foot to explore. How marvelous it felt just to walk.

I knew of Nome as the end of the Iditarod, that grueling winter dog-sled race beginning in Anchorage, which traces over one thousand miles north and west over roadless, empty land. Come spring and summer, though, Nome was not filled with dog mushers. Tourists, the usual Alaskan sort, walked the streets, some staging for an off-the-grid fishing or camping trip. Some looked like gold hunters. And then there were the birders, easy to spot because we never removed our binoculars from around our necks. An interesting bird could fly over in any location, at any time.

For a birder Nome consists of three long dirt roads that lead out of town: the Teller Road (which leads to Teller), the Council Road (which leads to Council), and the Nome–Taylor Road, known to all as the Kougarok, which leads eighty miles to nowhere. It was out the Kougarok where we were to find the Bristle-thighed Curlew.

After Gambell, Nome felt big, with roads and stores, bars, a Subway (where we would eat a few too many of our meals). Nome also seemed dominantly white, with the swaggering feel of a Western town. It didn't take me long to walk the few streets, along the way learning that Nome was, in 1899, a gold boomtown with five thousand greedy inhabitants. A sign marks where the red-light district operated, explaining in raised bronze letters that prostitutes paid a monthly fine of ten dollars, which kept them out of jail. I marveled that in modern Nome, so close and yet so far from the petroleum-rich north slope, gas ran at $5.94 per gallon. A pickup truck was the only vehicle to drive.

Midtown, I found the bust of the polar explorer Roald Amundsen. Stoic, unlovable Amundsen. There is no greater explorer: among many extraordinary voyages he claims first to the South Pole, first through the Northwest Passage, second through the Northeast Passage, and first over the North Pole. Nome is where he ended his 1906 voyage through the Northwest Passage.

The sculptor had perfectly captured the Norwegian's prominent nose and droopy eyes. I snapped photos, feeling it wrong that a glowing red sign for Verizon served as the background.

Finding the bust of Amundsen in Nome made me think about his heroic journeys, his determination and focus to go out again and again though safe return was never assured, even often unlikely. So it seemed fitting that the bird I most wanted to see, the Curlew, was also mythic, at least to me. It's more superbird than bird. It breeds in two distinct areas of northwest Alaska so remote that researchers did not locate a nest until 1948. But more than this isolated life, what awed me is that after the young fledge, they fly *nonstop*, up to six thousand kilometers over open water, to the Hawaiian Islands to winter. I like to imagine that initial push into the air, the first wing beat launching out on a flight of such a distance. What would it take to have the courage to commit to a journey so long and dangerous? Is the Bristle-thighed Curlew aware of all that lies in front of it: wind or hail, rogue waves or predators? Is it aware that safe return is unlikely?

To find such a creature wandering the wilderness outside of Nome felt as impossible as skiing to the South Pole. Our guide tried to calm our expectations as he reminded us again and again that the bird was not a given. If we failed in our search, we should not cry. If we found the bird, we should feel extra lucky.

On Bristle-thighed Curlew day, we rose at five. In semidarkness Peter showered as he did every morning, washed his thinning hair, trimmed his short red beard. He dressed as he did every day: neat khaki pants (amazing how they had stayed pressed during our travels), a button-down shirt, baseball cap to shield his bespectacled eyes. He was ready before me, sitting by the door, cradling his camera. I moved around the sterile hotel room making sure I had layers to transition from the morning temperatures in the low forties to the midday temperatures around fifty, to checking that the sun lotion was in my knapsack, to wanting to wear something special as if going on a date. But all I had was what I thought of as my birding uniform: loose-fitting beige pants that had pockets in unhelpful places, a long-sleeved cotton shirt, a heavy pile sweater, and a windbreaker.

"You have water?" I asked.

He shook his head, and I handed him a bottle.

We loaded into the van with eight others, knowing we would not be back in our hotel room until fifteen hours later. For the birds, I would endure being jounced about in a van for ten hours, dust pouring in through the cracks. As we drove, I tied a bandana around my nose and mouth to keep from inhaling too much dust, which meant I wasn't able to participate much in the conversations among our group that roamed from places birded (have you been to Texas? Florida? Arizona?) to birds seen (where did you get your Great Gray Owl?) to—surprise!—someone wanted to discuss pornography. I *had* read *Fifty Shades of Grey* (perfect airport reading) on the trip but wasn't going to admit to that, not because I didn't want to explain why I found the book an oddly tame good-girl's fantasy but because I knew I'd sound like a snob when I argued that the book is poorly written. Who reads soft porn for the prose?

Our unfortunate guide at the wheel of the rental van had to discern over all the disconnected conversations what was a real bird, one worth stopping for, or was that *just another* Long-tailed Jaeger flying like a winged spear, almost outracing the van; or *just another* Willow Ptarmigan squatting like a loaf of whole grain bread by the side of the road letting out its absurd song, *heil, heil, heil.* In eighty miles, we counted eighteen, the state bird perfectly abundant.

All of the couples sat side by side in the van, sharing water bottles and concern. "Are you comfortable?"

"Did you see the Short-eared Owl?"

"Are you hungry?"

Peter sat at the front of the van, the prized spot in that he could see the wide expanse we drove into and could easily lean out the window to capture a photo. Up front, he also avoided both dust and our random conversations. This gave a greater sense to our disconnect, the two of us on this trip alone together. Every evening while in Nome we returned to our hotel room so tired we hardly spoke to each other. And yet, elated by the land and the birds, I didn't let our silence make me sad.

Midway to where we hoped to see the Bristle-thighed Curlew, we pulled over to peer through scopes at a Gyrfalcon nest high on a cliff. A speck of head of this formidable falcon poked out of a scant pile of sticks over one hundred feet up a sheer cliff. From its austere perch, the bird had a long view down the valley, no doubt the perfect perspective from which to find a ptarmigan dinner. I hoped to see the Gyr drop from the cliff and bottlerocket out over the valley. Despite my wishes, it remained a dot in the distance.

From time to time we passed a house, bright yellow or blue, paint chipping, or a bare wooden box designed to survive. They gave scale to the place, the house like a jellybean left on a gym floor. I hoped that someone came in July or August to open windows, sit on the porch, and wave to passing trucks and vans. But for now, they sat empty.

As we jounced down the road, our guide told stories. He was a great storyteller, not a detail left out, and had plenty to tell about past tours, mistakes made, disasters averted, grizzly bears encountered (maybe guided trips were not all as sanitized as I imagined). As we drove toward the Curlew he helped to build the excitement by recounting a past bid to see the bird.

The terrain where the Curlews breed is rough, the earth bunching up in tufts called tussocks. We had already encountered these tussocks on our first day outside of Nome, so I knew they were hard to navigate. Tussocks are made of dirt and grass and are formed like mushrooms, narrow at the bottom and wide at the top, a grassy canopy on which you can step. At times they move, shift underfoot, making for difficult forward motion. At times your foot can slip off of the tussock, and like a sudden trap door, you sink an extra foot to the ground.

A woman on a previous trip stepped on a tussock that sent her tumbling. When she settled on the ground, she found that her ankle had snapped. The guide drove to the nearest cluster of houses, some ten or fifteen or thirty miles away. Though the sun still lit the sky, it was late at night. He banged on the door, waking everyone, to ask if he could borrow an ATV to rescue his client. They were all a little drunk—I might be making up this detail, but it fits the story—yet still game for the rescue. Off everyone careened on the ATVs.

In all of the empty tundra, the guide realized he could not accurately remember where the woman had fallen. The sun tilted toward the horizon becoming darkish as it does up north. They combed the wide field for hours; as it started to get lightish, they found her half asleep, in pain, certainly hungry and thirsty. They hoisted her up onto the ATV to transport her on a rocky, unpleasant ride out.

This story, which took over forty-five minutes to tell, and which I may have gotten only the gist of correct, served as an elixir; I became delirious at the prospect of searching for the Curlew. At the same time, I was concerned about some in our group—slightly reconfigured, thank goodness minus Bonnie of the Biglist—who had difficulty walking even on flat surfaces.

We poured out of the van, scopes and cameras at the ready as if we might be on a military maneuver. Fifty feet up the hill I wobbled on a tuft. My arm, set out to brace my fall, slid between two of those tussocks, sailing onward to land firmly on nothing. My fall stopped when my cheek smacked against the sharp short grass of the high north. My neck canted at an unpleasant angle.

I let my body settle, the initial shock of pain rippling through my neck, down my arms. The surprise of falling faded as I realized everything was intact, the pain dissipating, though I knew I'd have a sore neck for a few days. I looked over to see if Peter might help me up, but he was already a hundred feet up the hill, unaware of my struggle to right myself. I lay there for a quiet, surreal moment, half laughing that I had been worried about the old ladies in our group, and there I was the one on the ground stunned by the view of the vast land that stretched to infinity.

The season after my ballooning job, I enrolled in classes at the Institute Catholique in Paris. I thought of it as a movement from the profane to the holy, from champagne evenings to cold nights reading the Vatican II papers. While in school I sat side by side with future priests reading the New Testament and seeking out words I did not know in either language.

I am not Catholic, was raised with no religious tradition. But something about all of the cathedrals in France—where I spent a lot of time sitting and thinking—spoke to me of the power of this religion, and I wanted to know more. Beyond my class readings in the Bible or ecclesiology, I read stories of nuns and of nuns who had leapt over the wall. I immersed myself in the teachings of Dorothy Day and Thomas Merton. And one long weekend I walked from Paris to Chartres, the first part of one of the pilgrimage routes to Compostela at the tip of Spain. I walked with four thousand others, none of whom I knew.

The last morning of our walk I rose at 4:00 a.m. after having spent the night in a barn. I was handed a candle, lit it, and made my steps following the hundreds in front of me. We framed a farmer's field, dots of light moving slowly toward Chartres. And then someone began to sing a hymn, the voice low, joined by others. It wasn't a hymn I knew so I hummed, a vibration in my chest. The sound swelled, as if rising from the earth itself. Tears clouded my eyes as I moved forward in the near dark.

This moment in a field in France in the fall of 1983 is one that I hold close. It's not one I find easy to write about, words inadequate to the experience. The best I can do is to say that I have never so loved the earth.

There, lying in that field in Alaska, I felt again as if I heard a hymn rising from the earth. It was a low hum, no words, a vibration that I felt more than heard. And suddenly our guided trip made sense. I had traveled all this distance to press my ear to the ground, to hear the earth sing, to remember that the earth sings.

I hoisted myself up from my moment of contemplation, not wanting to fall far behind the group and then make them wait for me. I could see our guide a ways up the hill, a long-lens camera as well as expensive playback equipment strapped about his shoulders like a cameraman in a Hollywood movie. Over the other shoulder was slung a high-powered scope. Under the weight of all this gear, he led the charge, stopping from time to time to play the tape of the bird's song.

The most common call of the Bristle-thighed is a basic *chiu-eet*, which is why the Yup'ik Eskimo call them the Chiuit. I wished that this was the bird's name. Why name a bird for a feature—those bristled thighs—that no one ever sees and one that scientists do not yet understand.

The Curlew's song, however, is one that has been described in all of its layered complexity. Nuptial vocalizations include "Sweeping portamentos, dramatic frequency spikes, and multiple patterns of sequential frequency shifts." Additionally, there is the low whistle "similar to warning beeps of dump trucks in reverse" and the complex whistle and the whine. The "most typical whine [is] a wavering, plaintive whistle resembling *whee-ooo-whee-ooo-whee*." I wanted to hear the bird as much as I wanted to see it.

We spread out across the field, dots in a treeless landscape. Small vibrant flowers, poppies and lupine in purples and yellows, had been coaxed forth from the recently thawed earth. My favorite flower, the hoary lousewort, pointed skyward like a small fluffy sparkler. As I stumbled my way across that limitless field, I experienced the quiet elation of empty land.

Empty land is what the Curlew seems to prefer. When it's not in the expanse of Alaska in summer, it spends its winter on islands in Oceania and can be seen on beaches on Caroline Atoll or on lawns on Midway Atoll. Among the early reports of Curlew life and behavior, there's a lovely snapshot from 1891. At night, a caretaker on Laysan Island hung a cloth outside his door. In the mornings, he shook out the cloth; a flock of Curlews would run up catching the moths as they fluttered off. The land I looked out over did not hold a flock of these birds. Maybe one or two were out there, blending into the land, speckled on speckled.

"There," I said pointing into the far distance. The thrill of discovery jolted through me as I made out a slim shorebird on the horizon. Our guide rallied to my sighting, asking me what I saw and where. Being able to guide others to what you see in the field was, I had learned, an art. "Over there" doesn't work. "Top of the Beech tree" does. Too much detail (where the two trees form a V, the fifth branch on the left, the one that points toward the sky) distracts people, and some people like Peter simply can't locate a bird using the clock (bird at three o'clock). What I realized as I struggled to get the guide on the bird was that this land offered absolutely no point of reference, no bush or tree to orient us. I pointed in the direction of the bird, offering some distant hills as a line of sight. When that didn't work, I picked a stray cloud and dropped a line from its center to the bird. "See it?" I asked, hopeful. It took a while, the guide's frustration barely hidden, before the bird came into focus for him.

Had I really been the one to find the Curlew? It seemed improbably fun, a phenomenal find.

"A Whimbrel," he said promptly as he peered through his scope.

His sureness startled me as the birds are similar: medium-sized shorebirds with decurved bills. But I did not doubt his ID as his birding skills were awe-inspiring: all of the songs, chips, wing bars, length and color of tail feathers nested in his bones.

Disappointment washed through us all, then we got back to work.

We fanned out in a line, as if looking for a body, though we weren't looking down, we were looking out. We walked like this for a while until my eyes started to ache with trying to find the shape of a shorebird with a long bill. Another tour group appeared up the hill, a flank of humans all with the hope of a Curlew in their hearts. Though I recognized this gave us greater searching power, I wasn't happy at the reinforcement; I wanted all of these hundreds of acres to myself. Still, we were a puny bunch given the extent of the land we had to scour.

Two hours in, when our guide found the bird, it was ambling about amongst the tussocks. Its speckled narrow body blended in with the brown and green of the grasses. We kept at a distance, only able to make out details through the scope. Squinting through the powerful eyepiece, I took in the buffy chest, the prominent head stripe that ran from the bird's nearly invisible dark eye. What I lingered on was the bill, long and curved. The genus to which the Bristle-thighed Curlew belongs is *Numenius*. New moon. That seemed just right.

The weekend before we left on our month-long trip to Alaska, I had been invited to speak at Bard College's baccalaureate ceremony. Baccalaureate is the intimate moment of graduation, only students and faculty coming together to celebrate and say goodbye. In the quaint, dimly lit stone chapel, I stood in the narrow wooden pulpit, looking out at a sea of bright, exhausted faces. For that moment I was a preacher. And I preached the Bristle-thighed Curlew. I spoke to those creative students, describing the Curlew's long flight, it's steadfastness in that flight and urged them to lift off, to commit to their own lives. To launch out over the sea for miles and days would be hard with so many dangers, with so much unknown. But staying safe on shore was harder.

Of course like all such speeches, I was talking to myself as well. And part of me half believed that in bringing the Curlew to these students perhaps I was bringing the Curlew to myself, willing it into my life. Now, in this wide field outside of Nome, here it was, threading its way through the tussocks.

Getting to this bird had been so calculated—I had spoken of, dreamed of this bird, had paid a silly amount of money to a tour company, flown from New York to Anchorage to Nome, and then driven eighty miles on that awful dirt road. I appeared as determined as the person who creates a five-year plan then systematically works her way there. I am not that person.

Maybe I hadn't set long-term goals because what happens when you set a goal and don't reach it? Perhaps I wasn't strong enough to weather that devastation. But there was another downside to the goal, which I was only just realizing: wanting something and getting it. I now understood why it was so rare to find someone who had made her fortune who then sat back and enjoyed it. I was already, shockingly, hungry for the next bird. A mixture of excitement and longing swirling inside of me made it hard to linger in the satisfaction of seeing this extraordinary Curlew.

I looked over to check on how Peter was reacting. He had his camera to his eye, taking long-distance shots. Soon he lowered his camera, paused to give the tundra one more scan, then turned to walk downhill. I saw that he was headed to the next best thing. Part of me admired that, the way he stayed hungry for the next bird, a celebration of life. But his walking toward was also a walking away, and in this moment on that hill I had so loved outside of Nome, I saw that he was walking away from me.

For a while longer I absorbed the cold air, the exquisite northern light. Then I turned and walked down the hill, my heart set on the next wondrous bird.

A Perfect Fall Day

Greig Farm, Dutchess County, New York

Perfect fall days on the East Coast are a particular torture. There is no way to properly celebrate the clean sky, the feathered breeze, the gentle angle of the sun. The poignancy of these days makes every color, every gesture, every thought and emotion stand out.

To try to capture this day, Peter and I were walking Greig Farm on a Saturday late morning searching for sparrows. It had been three months since Peter and I had returned from our Alaska travels. Our last great find came at a campground off the Glenn Highway when we were done with our guided tour and were finally traveling alone in a rented RV. There, in a campground by the Tolsona River, we stumbled on a pair of Boreal Owl babies. Baby anything is a treat, and baby owls, all fluff and innocence, even more so. But Boreal Owls? It seemed a true miracle.

As thrilling as these owls and all of the Alaskan birds had been, I was content returning to my local birding haunts. The fields and marshes near home were my friends. Right away I noticed the ash trees that had fallen to the emerald ash borer on Cruger Island Road and admired the new beaver lodge under construction in the North Tivoli Bay. Hearing the Red-winged Blackbirds and Swamp Sparrows in the Bay made me feel welcomed home.

I would always crave the awe and thrill of places that overwhelm me like Alaska, and I would always need the intimacy of home to ground me. I saw that my bird life would involve my own migrations, taking on the outward energy of travel, followed by the sureness of nest building. For now, I felt like I needed the familiar for a long while, and Greig Farm was just that.

Greig Farm is wide, open, flat land, with views to the Catskill Mountains to the west. The birds never disappointed, and neither did the sunrises and sunsets that frame the horizon. The farmland shifts constantly, crops rotating while many acres lie fallow. Beyond the expected sparrows like Song and Savannah, Vesper and Grasshopper Sparrows also visit. In fall Snow Geese have landed by the hundreds, blanketing the fields with their noisy whiteness, while American Pipits step like old men through the furrows, and in winter Horned Larks fret. Sometimes Greig harbors a flock of Snow Buntings, doing a jouncey dance across the fields. Through early summer Kestrels work the open fields. Amidst this parade of interesting birds, rarities appear. Exactly a year before, Peter and I had found a Red Phalarope at Greig Farm.

On Red Phalarope day, puddles dotted the property from weeks of fall rain. Just as we were following a few pipet-sized birds, a flock of five shorebirds circled in front of us.

"Those are good birds," Peter said as we followed them through our bins. "Keep an eye on them." They circled and circled. "Land," Peter whispered, willing them to ground. Sure enough, they landed in a large puddle 150 feet in front of us. I stayed put with the scope, while Peter inched forward with his camera. Two Pectoral Sandpipers and a Least Sandpiper searched for a meal along the edge of the water. Next to these stepped a bird with a thin dark bill and tiny head.

Was it a Red Phalarope or a Red-necked Phalarope? Both species breed on tundra ponds in the Arctic and spend their winters over the open ocean. An inland farm was a long way from where either species belonged.

Peter and I had seen a Red-necked Phalarope in Seward, at the start of our Alaskan travels. But that earlier sighting did not help us with our current identification as our Alaskan bird was in its splendid breeding plumage, with a gray chest, white throat, and rust red neck. This bird before us had a straight black bill of medium length, beautiful gray-blue feathers, with a wash of red around the neck. Nothing stood out in its markings to make it easily identifiable as a Red or a Red-necked.

I remembered watching the Alaskan Red-necked Phalarope spin and spin as phalaropes do on a pond in order to lift food to the surface. Phalaropes are a species where the female is larger and more brightly colored, and in another reversal, she leaves the eggs for the male to incubate (while she pursues other males). Like when I learned of the Bicknell's Thrush's polyamorous ways, whenever a bird species defied the expected roles of plumage or egg and nestling tending, I felt an odd, admiring affection for them.

The Phalarope loitered, completely unaware it was about to become a celebrity. Peter posted to both the local and New York State lists asking for help with the ID. While top birders drove in that afternoon to see the bird, others scrutinized Peter's photos to agree that we had found a Red Phalarope.

The next morning, sharp-eyed birders descended on Greig Farm, driving from throughout New York as well as neighboring states. The goal was the Phalarope, but Greig is so rich in birds that someone stumbled on a Nelson's Sparrow, and when Peter returned to visit the Phalarope he located a LeConte's Sparrow, a first for Dutchess County. It was a Dutchess County version of the Patagonia Picnic Table, that legendary place in southwest Arizona where a birder, drawn in by a Rose-throated Becard, then found a colony of Five-striped Sparrows and, later, Thick-billed Kingbirds. News of these special birds spread. Soon people traveled to this innocuous roadside stop in Arizona to find state records of Yellow Grosbeak, Black-capped Gnatcatcher, and Yellow-green Vireo. From this the truth emerges that where there is one good bird, there are others. Or, another way to think of it is: there are always good birds.

In the last year, Peter and I had been on a roll with good birds. A week after we saw those Boreal Owl babies in Alaska, he sent me an email:

> I was just looking at photographs when I realized that it was one year ago on this date, July 9th, 2011 that we found a Henslow's Sparrow in Ames, New York. That got me thinking about what an amazing year it was for us finding rare birds. Think about it, between July and November we found a

Henslow's Sparrow, a Whimbrel, a Red Phalarope, a LeConte's Sparrow, and a Harris' Sparrow. It is really amazing. It would be a good year if you found any one of those birds, but to find five good rarities in five months!

It makes you realize that every time you go out, you hope to find some treasure. But mostly you find what you expect to find. When it does happen, when you do find the unexpected, it's almost shocking. How could there be two Boreal Owl fledglings sitting right in front of you—what are the odds! How do you explain the feeling when you get out of your car and you hear the "fleet" of a Henslow's Sparrow! I guess there are moments in life that are meant to be experienced. I am glad you can write about them.

And I was glad Peter took such beautiful photographs of them. And of course beyond the pleasure of the birds themselves was the sense we found them *because* we were together. We were a team, a good, even inspired team. The birds became a sign that we were blessed, meant to be, like the couple who raise talented, beautiful children. To our birding friends, we were the perfect couple, out every weekend, keen for the next bird. And yet there is always an inner story, the one only the couple knows.

"You know what the rumor is?" Peter asked as we strolled across the Greig Farm field.

I shook my head.

"I almost feel bad telling you this."

I watched a Northern Harrier float across the field, all focus and grace.

"That we were on our honeymoon in Alaska."

I almost laughed.

"Who started that?"

He shrugged.

I waved a hand dismissing the comment. The rumor and all it implied could have sunk me into a sullen mood. If a honeymoon is about sharing a beautiful place with the one you love, about building a store of happy memories that will solidify a relationship, then we had been on sort of a honeymoon. On our return from Alaska, though, it was clear we were further apart than ever. We only saw each other on weekends to bird. Gone was our Friday night dinner, the one couple-like thing that we did. If Alaska couldn't get us past where we had gotten stuck, if finding baby Boreal Owls couldn't bring us together, then we really were the best birding partners but perhaps not the best couple.

For the next hour, we combed the fertile land of Greig Farm.

"Savannah," Peter called.

"Vesper," I pointed, catching the white outer tail feathers.

"Nice."

Crows coursed through the sky, a river of birds that painted black on blue, like stark, clean line drawings

Side by side, we followed the mowed path through the grasses, stopping and starting, happy with the glimpses we had of small birds

"If we are just friends," Peter said, "let's call it that." He spoke so softly I had to lean in to hear him.

It was what I knew, even what I wanted, but still, I felt my heart drop.

"OK," I said.

We continued to walk as I wondered at how easy it was to acknowledge what we had become.

Peter took my hand, our fingers entwined solid, comforting as we walked back to our cars. As we hugged goodbye I breathed in relief at this bittersweet decision, made so simply while walking across the fields at Greig Farm looking for and finding Savannah and Vesper Sparrows.

SURVIVING THE WINTER

South Tivoli Bay, New York

The Snow Goose perched on a rock above the March-cold water of the Sawkill Creek. It shifted from one webbed foot to the other, as if it might be trying to keep itself warm. It had a few feathers sticking out of place, like someone had ruffled it up. I set up the scope so that my student Christina could get a better view of the bird and watched as she curved her tall frame, in order to peer through the eyepiece.

This Snow Goose blew in with Hurricane Sandy, which pummeled the East Coast beginning on October 27, 2012. It landed on the South Tivoli Bay, just off of the Hudson River one hundred miles north of Manhattan. As flocks of Snow Geese migrated through, it never joined its comrades en route from the Arctic heading south, perhaps to the Texas coast. I saw the bird through November and December and finally concluded when fall migration stopped that it was not going to leave; perhaps it could no longer fly.

I was sure this bird wouldn't make it through our northeast winter. Yet when I checked in December, there it floated by Buttock Island. Again in January, it held its own in a few areas of open water near the mouth of the Sawkill. The odds of surviving seemed slim when the coldest days of winter hit. The ice formed so hard and thick that iceboats skidded across the wide, smooth surface of the shallow South Tivoli Bay. Surely a bird couldn't find food on a frozen bay. Yet there it was, in early March, carrying on its goosey life.

I wanted to adopt the Snow Goose's nonchalant attitude. It had been a rough winter, and not just because of the extreme cold. Though our breakup had been easy, what followed post-Peter was not. We all know that if you love and are loved you are strong and willing to trust, like a trapeze artist who takes flight confident her partner will catch her. I only now realized that with Peter I had become weaker. I didn't blame him or myself. And we did love each other, but despite pushing hard into the forests and swamps, hoping for a good bird, we only saw Cardinals and Blue Jays. Chickadees. Good enough birds, but not what we both wanted.

The environmentalist writer Rachel Carson wrote: "Those who contemplate the beauty of the earth find reserves of strength that will endure as long as life lasts. There is something infinitely healing in the repeated refrains of nature—the assurance that dawn comes after night, and spring after winter." I was still waiting for dawn. Yet I knew that spring would follow, that in time I would return to myself—how many times had I done it before? The heart, after all, is a muscle, easily bruised but amazingly elastic. But birding on my own through the winter had brought on a loneliness I had never known before. My emotions jumbled in unfamiliar ways, not the usual ache of longing or fog of sadness that shrouds a heart in grief. This felt deeper, more physical.

From time to time Peter and I met on a Saturday morning to bird. But I found I didn't enjoy our time together in the field, as he was often distracted by texts and would leave midmorning to head into the rest of his day. I understood he had things to do; he had a new girlfriend, or rather an old girlfriend from his early twenties, the true love he had always wanted to marry, returned to finally get it right.

"She couldn't find a bird if it was fifteen feet in front of her," he joked when he told me about her. Comforted that she was not his birding partner, I also had to accept that neither was I.

Aldo Leopold, in his beautiful essay "Marshland Elegy," writes: "Thus always does history, whether of marsh or market place, end in paradox. The ultimate value in these marshes is wildness, and the crane is wildness incarnate. But all conservation of wildness is self-defeating, for to cherish we must see and fondle, and when enough have seen and fondled, there is no wilderness left to cherish." This is the dilemma I faced: I wanted to bring my students to the birds, to experience what I find most exhilarating about life and come to love and maybe even want to protect the natural world. But I also worried that in bringing my students to the natural world I would be giving up great secrets. And that these secrets, once fondled, would lose their power, or even be destroyed. My worry wasn't all speculation. In one nature writing class, I took my students by canoe to a dollop of an island in the South Tivoli Bay, only to have two students later return to the island and set it on fire.

But I didn't need to worry about a stampede of students in the woods, scaring off the birds of the Bay. Only one student showed up for my Wednesday morning walks: Christina.

So on this cold day in March 2012, it was just the two of us admiring the Snow Goose. We then turned down the trail, passed the field station, and entered the hardwood forest, where the trees had not yet leafed out. Everything was emerging from the winter freeze, the ground damp, ready to welcome the delicious spring. With glimpses into the wide, shallow South Tivoli Bay, Christina followed close behind me, like a gosling. Five foot ten, broad shouldered, with a ponytail that ran to midwaist, Christina had to work to contain her exuberance. I had introduced her to a Yellow Warbler in a class the previous fall, and that was it—spark!—she was hooked.

After twenty minutes of walking the narrow packed trail, Christina and I arrived at Buttock Island, which is a promontory with two mounds of eroded dirt, cracked with a narrow trail. Spiky water chestnut seedpods—devil's heads—rimmed the trail, like thousands of black dice tossed on shore. The end of the small promontory offered a rich perch from which to scan the wide Bay.

I spotted Bufflehead through my binoculars, then set up the scope to get a closer view. Christina was so new to birding she didn't yet have binoculars. She peered through the scope to admire the male Buffleheads with their white bodies and black capes, which makes it look like they are wearing a white half-helmet on their small black heads. She stepped back from the scope, her hand covering her mouth as if to keep herself from crying out her pleasure. Watching Christina, I relived my first thrilling months birding, the exquisite sense that the world was so much brighter and more astonishing than I had ever known. I was envious that she was learning this in her twenties, when all of the little birdy details land and stick like Velcro.

We scanned the Bay, finding Great Blue Heron wading in the shallow water, Mallard and Black Ducks quacking about, and Common Mergansers scooting along the far edge of the Bay. A Bald Eagle soared overhead, putting up a flock of Ring-billed Gulls. The smell of thawing ground, dense like dark coffee grounds, filled my senses. Soon the world would be alive with bird song and the buzz of insect life, especially the whir of the seventeen-year cicadas. For now, stillness, until a train screamed south across the breakwater, heading toward the city.

"OK, time to get to work," I said, satisfied. I gathered up the scope.

For me, work meant reading student papers, answering emails, juggling the work of a writing professor. For Christina getting to work meant going to a studio in the art building and drawing birds. Birds and art go together.

The number of famous ornithologists who were also artists is impressive: Alexander Wilson, John James Audubon, Louis Agassiz Fuertes, Roger Tory Peterson, and David Allen Sibley to name a few. But, again, my list is missing the names of women ornithologist artists. Most have historically been less known, though thankfully that is changing. I am thinking of biologist, artist, and author Sophie Webb, whose work ranges from watercolor to ink drawing to bird guide plates (for the *Birds of Mexico*). And there are many avid birder artists such as Catherine Hamilton, who has traveled the world and draws the rare and endangered, and Julie Zikafoose, whose book *Baby Bird* is an artistic and ornithological beauty. These are just three in what is a longer list of contemporary women bird artists.

Does the birder become an artist? Or is it the other way around, that the artist finds her subject? For Christina it was the latter. She had been painting since she was a kid, and she had from the start been drawing from nature. But taking birds as her subject happened in college. Every day she painted a bird she had recently seen, as lifelike as possible. These were in a small notebook, the birds accurate. I saw that her technical ability to reproduce on the page what she had seen in the field pushed her attention to details of color or bill shape; her skill as a birder went hand in hand with her skill as an artist. Beyond these meticulous drawings, there were her large paintings where, in broad strokes of color, she captured both the spirit of the bird and her enthusiasm for the encounter.

Christina regularly reported from the field, sending me photographs of birds she couldn't identify (a White-eyed Vireo!) or texts alerting me to new birds in the Bays (Horned Grebes in the South Bay!) and emails about her observations: "They can JUMP!" A Turkey Vulture hopping over another in order to get at some food made her ecstatic. And through her reports, I, too, learned more about the birds.

Christina spent hours near the Bard College dump where Black and Turkey Vultures, often called buzzards, waited, brooding over the next meal, or circled like hang gliders playful in the thermals. "Notice how when they soar they teeter like they are drunk," I coached, hearing in my words and description Peter's voice. (Florence Merriam Bailey described the Turkey Vulture as wobbling "from one side to the other like a cork on rough water.") Christina's final paper for my class reveled in Turkey Vulture-ness that included a link to a marvelous video of the Buzzard Lope. I watched, entranced, as a man shuffled, crouched with outstretched arms, and then hopped buzzard-like over his carrion, a white handkerchief, to the song, "Throw Me Anywhere, Lord."

That Christina so loved the vultures—a bird, I admit, I hadn't spent much time admiring—impressed me. The Turkey Vulture is not the most loveable bird out there. From the start, they were seen as lazy and unwanted. The great naturalist Georges-Louis Leclerc, Comte de Buffon, describes them as "voracious, slothful, offensive and hateful, and, like the wolves, are as noxious during their life, as useless after their death." Even Florence Merriam Bailey, who celebrated most birds, hated them. "They were grotesque birds. I often saw them walk with their wings held out at their sides as if cooling themselves, and the unbird-like attitude together with the horrid appearance of their red skinny heads made them seem more like harpies." The vulture is somewhat reclaimed from harpie-dom by William Faulkner. "You know that if I were reincarnated, I'd want to come back a buzzard. Nothing hates him or envies him or wants him or needs him. He is never bothered or in danger, and he can eat anything."

Faulkner isn't entirely correct; buzzards don't eat everything (carrion is not everything), and we do need them in our ecosystem—eating that carrion is a great service. What Faulkner gets right, and exquisitely so, is the aloof independence of the vulture.

When I found Christina creating a papier-mâché, oversized Turkey Vulture to perch on a branch in one of her art installations, I knew I had found someone just wacky enough to become a regular, and fun, birding partner.

On our return from our walk to Buttock Island, Christina and I stopped on the wide lawn below the blocky, white Blithewood mansion that keeps Bard students believing in ghosts.

"Hang on," I said, peering north into the Bay. "Those birds at the far end aren't Mergs. They have dark chests."

Christina squinted through the scope. "You're right, they are different."

I could see that she expected me to know what the ducks were. It both pleased and unnerved me how she counted on me to identify what we saw and heard because there is nothing like teaching someone else to highlight what you don't know.

Those first walks with Christina, I often heard myself say: "I don't know what I am doing."

True, I still had so much to learn. But I also knew that claiming I didn't know anything was too often a part of the narrative of women, especially in the outdoors. Too many times I have heard or read: "I didn't know what I was doing." Most people don't know what they are doing—at times that is why we are doing it, for the new experience, whether for the thrill or to learn. But women admit to not knowing, at times celebrate it. There's something refreshing in that honesty, but I worry, too, if it's a way of putting ourselves down, of not claiming our strengths.

Even though Christina and I left a lot of puzzles in the field, I gave up saying "I don't know what I'm doing." I knew what I was doing: I was looking for birds.

And now we had found some interesting ducks.

"Look up Canvasback," I said. We both peered at the guide. What floated on the water matched the image in the book: a large duck with a white body, black chest, and a dark red head. It had a big black doorstop of a bill. We both grinned like lunatics.

"I'm sorry, but we have to get a closer look," I said. This was too exciting a find to have only a long-distance view. Now I was late getting to my office, to answer emails and meet with students.

Christina and I retraced our steps, then continued past Buttock Island. We left the trail to navigate through the leaf-bare trees along the edge of the Bay, creating our own path, or following one set by deer. It wasn't easy going, bushwhacking our way closer to the Cans. We approached as quietly as possible, dry leaves crunching underfoot.

Alexander Wilson, approximately two hundred years before us, also had his Canvasback moment:

> Slow round an opening we softly steal,
> Where four large ducks in playful circles wheel;
> The far-famed canvass-backs at once we know
> Their broad flat bodies wrapt in penciled snow;
> The burnished chestnut o'er their necks then shone,
> Spread deepening round each breast a sable zone.

Not surprisingly, Wilson is not remembered for his poems (though in his day, he, like John Burroughs after him, was a huge success). Perhaps it is that this poem is more of a bird ID than a poetic tribute to the bird. Would Wilson be a household name like Audubon had he completed his *American Ornithology*? He only finished eight of ten volumes, dying in 1813 at age forty-seven of dysentery. Perhaps it was a mistake to drink water directly from the Ohio. Images of Wilson show a dark-haired man with high, thin eyebrows and melancholy eyes. Perhaps that somber look was because he left more unfinished; among his letters are tender ones to a young woman named Sarah Miller.

Do we all die with regrets? When my father's heart neared its final beat was he, crunched in his car, thinking: *damn, wish I'd finished writing that book*? I doubt it. It's those left behind who hold regrets. I wish I had that completed novel, to keep my father close, even in this small way. I wish, too, I had Wilson's complete *American Ornithology*.

"Now I can die happy," or "now my life is complete." I found myself repeating these lines in the field after seeing a particularly pleasing bird. It was sort of a joke, but not entirely. Not that I was anticipating my actual demise. But I did feel like the birds made me complete, gave me something I had wanted my whole life, without knowing that's what I wanted. I just had always felt out of sorts, restless, hungry. Now that hunger translated into getting up early to find birds. I felt good in my skin, as the French would say. Settled, not in a boring way, but settled in an active, independent way. Settled in a focused way.

John Burroughs wrote that in order to see a bird you had to hold it in your heart; without love your quest will fail. That meant that in all of my hours, days, months in the field learning the birds, holding the Brown Thrasher or the Least Bittern in my heart in hopes of seeing them, I was also learning love. All of that emotional exercise meant that, thanks to the birds, my heart was more generous and flexible. I counted on this new bigger heart to help me to heal, as living in this post-Peter world left me dizzy with daily emotional contradictions. At once I was more content, more fully in love with this life and ready to share that love, while at the same time I experienced a deep hollow. The collision inside of me left me at times euphoric, at times flattened.

While Christina painted she listened to a tape titled "Who Cooks for Poor Sam Peabody." The narrator, who sounds like she is twelve years old, offers the bird song then a mnemonic to help remember those songs. In the field Christina would sing, "look up, over-here, see-me, up here," for the Red-eyed Vireo. Or "little old ladies don't chew chew" for the Louisiana Waterthrush. She also adopted my shorthand ways of describing a call. In the field she would point, "Breepers" (Great Crested Flycatcher) or send me texts, "*bizz buzz* on Kidd Lane." I knew she meant Blue-winged Warbler. Passing on what I knew to Christina was pure fun, because never has anyone had a more enthusiastic student.

With Christina, I experienced how details of birds, how we speak of them or remember them, get passed down from one birder to the next, one generation to the next. I kept hearing Peter's voice disguised as my own as I explained to Christina that the falcons were professional killers or that a proper birding outing required her to be up at five in the morning.

I was sure she would not be able to get out of bed to join me in the field. But for a week Christina systematically set her clock earlier and earlier until she rose at five. She waited at the corner, ready to go when I swung by at 5:30. We had taken to going out not just those Wednesdays before class, but Fridays as well. Since I didn't teach Fridays, our mornings in the field soon extended until they often took up the whole day. Christina at my side, I reached a beautiful bird exhaustion that often found us with the sun going down sitting in the parking lot of a supermarket devouring a roast chicken, barbarian-like, with our hands. "Eat bird to see bird," Christina would laugh, mouth full.

After every outing Christina sent me a text: "that was epic!" For her, even the quiet days were an outstanding event, every bird a glee-inducing treasure.

And now on this March walk, we had these Canvasbacks. Christina and I crept through the woods, snagging on bushes and stumbling over fallen trees. I wanted to get to the water's edge for a clear view of the Canvasbacks.

Cans are gregarious, often traveling in large rafts of up to five hundred. In February 1991, a birder reported 2,500 Canvasbacks in Tivoli Bays. I tried to imagine the Bay blanketed with the large ducks. In his *American Ornithology*, Wilson described these big ducks, which he calls the "Canvas-back Duck," arriving on the Susquehanna River in the fall in such numbers that they "cover several acres of the river, and, when they rise suddenly, produce a noise resembling thunder." He also offered advice on the best way to shoot them: at night, when they huddle in together. In one instance, gunners shot up to 240 in a day, selling them for twelve and a half cents apiece.

Despite hunting and environmental challenges, Canvasback populations, though fluctuating since the 1960s, are not in one of those all-too-common free falls. Still, they are one of the least abundant of our North American ducks. These few on the South Tivoli Bay were a rare opportunity for me to take in the Canvasback beauty as these birds were passing through the Hudson Valley on the way north to breed. They were not numerous, and they would not stick around long.

Christina was now late for class. As one of her professors, I recognized I was encouraging irresponsible behavior. But I couldn't stop myself. A spring duck fever gripped me. For a moment I thought it wrong to take

Christina with me on this caper, until I realized from the glow in her eye that she had been taken by her own duck ecstasy.

"I'm sure my teacher will understand if I tell her we found Canvasback," she said.

"Just don't mention my name, whatever excuse you come up with." We both laughed.

Moving on, we stopped whenever we had a peek through the leafless but still dense invasive honeysuckle and prickly multiflora rose. Wilson named the duck *Aythya valisineria* after the wild water celery *vallisneria* (Wilson misspelled *vallisneria*), which the duck feeds on. That is what the half dozen Canvasbacks were doing, foraging for the silky grasses shoulder to shoulder with the more common Mallards and Black Ducks. We whispered our excitement over what was so simply beautiful: dabbling, grazing, floating with your flock.

#1 Birder

Dutchess County, New York

On a sunny Sunday afternoon in April, I called my sister Becky just as she was finishing dinner in Paris. Her report on life in the last week rang of success. She had organized a conference on the history of education; her daughter Alice earned the highest grade in her law class. I had my own good news.

"I'm the number one birder in Dutchess County."

Silence shut down the wires between the Hudson Valley and Paris.

"Who decides this?" she asked cautiously.

I laughed, an admission that this ranking had nothing official about it. My claim originated from eBird, a site launched in 2002 by the Cornell Lab of Ornithology along with the National Audubon Society.

eBird puts to work all of the birders in the field. We go out, see the birds, then log in what we have seen and where. With thousands of ears and eyes in the field, researchers have access to an enormous bank of information about bird movement, distribution, and abundance. There are a lot of things that scientists do with eBird data. One of my favorites is the creation of heat maps, or occurrence maps, which show the move-ment of a particular species as it migrates. These maps make the mass of birds, marked as orange dots, look like a fiery ball in various stages of brightness. Studying them, you can almost hear the whoosh of birds as they move overhead.

These heat maps are not just great to anticipate the arrival of a particu-lar favorite, like the Black-throated Green Warbler or a Brown Thrasher working its musical medley. The maps are used in conservation projects like Bird Returns, run out of the Nature Conservancy. With these maps, researchers can pinpoint when shorebirds are going to need habitat on migration from South America to the Arctic. At just the right moment, they pay farmers in California to flood their rice fields, creating "pop up habitats." In its first year, 2014, forty-five farmers collaborated, produc-ing 15,000 acres of habitat for birds. This was a huge success to have such participation in the project, given that California was in the midst of a drought crisis.

But many birders in the field are not thinking about producing data for research or conservation. It's about the list. Every time you enter the birds you have seen, eBird keeps track of your sightings and numbers for the year, or month, or for an area. I already knew that I wasn't capable of keeping a life list—to this day, I can't say how many birds I have seen—and that I found the urge to list off-putting. But, if eBird didn't prime my competitive juices, it did induce a special bird envy. When I got my "needs list" for Dutchess County and saw that Deb Tracy, driving around with her wee dog, had found yet another Screech Owl sunning in a hidey-hole, I wanted to see it as well. Better yet: I wanted to be the one to have found it.

The desire for a big list or envy over another's sightings—eBird has tapped into these very human competitive emotions. With a click you can see the top birders in the county, state, or country. You can track who is doing what and where. So everyone birds more actively, which means more data for those using this information. It's perfectly ingenious, one of the few fields of research where you have people working for you tire-lessly, daily, dutifully, and for free.

When I first heard about eBird, my local bird people still communi-cated through a Yahoo Listserv or by phone. People on walks would ask: "Do you eBird?" as if asking about something slightly illicit. Everyone shook their heads and said no, no they wouldn't eBird (*just say no*). Then, so fast, everyone was eBirding, and it was obvious that everyone was birding more often, and in a more focused way. Peter stopped birding outside of Ulster County. His county eBird list became his obsession. If I wanted to bird with Peter, I had to leave Dutchess County, cross the Hudson River, and enter Ulster, his county.

My whole life I have tried to avoid competition. In high school, I thought I should follow in Becky's footsteps as a runner—she competed nationally in high school and later she captained the Harvard track and cross-country teams. When I showed up at track practice the coach beamed, "Oh, you are Becky Rogers's little sister." I should have been good. But after a few weeks, during which I frequently doubled over sick, I quit. I didn't have the lungs, the discipline, or the nerve to compete. I was grateful when I found rock climbing where overt competition did not yet exist.

In the small world of rock climbing in the 1970s people were, of course, competitive. But there were yet no formal competitions. I reveled in the after-climbing talk, where we discussed over beers who had done which route and how. These tales were often filled with high jinks and not-so-subtle bragging. Out of these tales local heroes emerged while cheaters and liars were shamed. This informal competition was soon to change.

When I was in high school, a rumor spread that the Russians were speed climbing. Not to be outdone by the *Russians*—this was the 1970s, so we were living deep in the shadow of the Cold War—my climbing partner Neil and I went to our tiny local cliffs, set up a rope, and started the stopwatch. It was an odd day, filled with falls and laughter. I knew in an actual competition I would not do well. It probably was no coincidence that once climbers started training and competing on indoor climbing walls my passion for climbing began to fade.

Now here was birding, an activity—not even a sport—for ladies in floppy hats, and yet it was so full of competition. The introduction of eBird into the birding world was like the advent of official rock-climbing competitions: birders jumped on it with a zest that was seductive.

For me to compete, however, was absurd. When I started birding, my goals were modest as I knew I had started too late in life to be anything but a decent birder. What I could do was compensate by being loyal to the birds: going out daily, paying attention to the ordinary while seeking the rare. I imagined this was like practicing the piano every day. I had my scales down, but I would never be able to play a Bach two-part invention expressively. eBirding encouraged this approach, as I logged in my daily efforts, kept track of the flow of birds through my life. It was the bird version of the journal I have kept since I was a child. And eBird tracked the numbers for me.

So there I was in April 2013, number one.

The difficulty with measuring one's ability through eBird is that some of the best birders don't eBird. It also happens that birders make mistakes, log in birds they thought they saw but misidentified (which I have done more often than I want to think about). So, though a good birder is often at the top of the eBird list, it's also possible to be number one and not be anywhere near one of the better birders in an area. That was the current situation I was bragging about.

"The pressure is on," I said, sighing into the phone. "Everyone is rooting for me." By everyone, I meant Mark, now the president of the Burroughs club where I had started my bird journey. And Peter. We were working our way toward a warm but cautious friendship, one where he continued to mentor my birding, if most often from a distance.

"Nice job getting a woodcock so early," Mark emailed me. "Pine Warblers should be back on Cruger Island Road," he coached. And "There's a Wigeon on a pond off of Route 9." He knew I needed a Wigeon. Peter urged me to think big. "There's no reason you shouldn't have a King Rail in the North Tivoli Bay." It was Peter who put it into my imagination that I could be number one when he wrote, "Look out Adrienne [then ranked #1] Susie Rogers is gunning for you." In a flash, I thought, *why not?* It took about two weeks of focusing on what birds I could add to my list. When I hit number one Peter sent an email that read: "#1#1#1#1#1#1!!!—from, #3."

In this push to number one I did not lose my mind and soul to games and ego. What I realized was that pushing for a bigger list had interesting benefits as I focused on birds in a new way. I thought about what birds should be in my area, and when, and how I might find them. So when I noted a Prairie Warbler missing from my list, I walked the fields near the North Bay that I did not usually frequent to hear the ascending song, which Roger Tory Peterson described as similar to "a mouse with a toothache." When I realized I didn't have a Golden Eagle, I trekked out to the Thompson Pond area (without luck). The absence of a Peregrine Falcon made me fret. I then proceeded to spend a substantial, some might say silly, amount of time crouched near the train tracks peering hopefully up at the Kingston Rhinecliff Bridge where the birds often nest. eBird made me think more accurately about bird movement, habitat, and behavior. It also made me think about my own behavior.

I imagined that Becky was hoping that my birding habits would mellow once Peter and I separated. Instead, my birding had taken this odd, semicompetitive turn.

"I'm exhausted," I joked. "Being number one is a heavy responsibility. When I'm away next weekend I'm sure to fall behind. The stress of that is enormous."

"The problem is that you're half serious," she replied.

In all of this, I had Christina at my side. Shin splints had sidelined her from training with Bard's track team, so she had more time to bird. On our outings Christina wore black half-tights (to help with her lower leg injuries) and black neoprene toe shoes. At first I worried that this smart, kind, nerdy young woman was making herself even nerdier with the birds. But soon she had binoculars around her neck and a long-lens camera dangling from her shoulder. She cut off her hair for a tomboy look, sewed images of vultures onto her jacket, and moved with a vulture's swagger. I sensed the tattoos were soon to follow. Before my eyes, she was transforming into what I would call bird cool.

Because we were focused on our success in Dutchess County, together we got to know every pond and wooded warbler hangout across 825 square miles. "Why didn't I know about this place before?" I asked Christina as we left Peach Hill, the warblers dripping from the apple trees. We spent a lot of time at Thompson Pond, a wide shallow body of water filled with ducks and geese, near Pine Plains, and Buttercup, an Audubon Sanctuary. One late afternoon we walked the rail-trail north of Amenia to find a Common Gallinule only to realize the rich bird possibilities of the swamp that borders the paved trail. All of these beautiful spots reinforced my feeling that I didn't have to travel distances, like to Alaska, to see interesting birds. It just took driving down unexplored roads and walking unused paths in the land around me.

Dutchess County is a great place to become obsessed with listing, as bird records in the county date back to the 1870s. The first county-based Christmas Bird Count was in 1901, only the second year of the CBC nationwide; a bird census has been conducted consistently in May since 1919. There are also extensive migration, nesting, and other records kept since 1885. Much of this data was collected by a man named Maunsell Crosby.

Crosby, born on February 14, 1887, lived most of his life on the family property, Grasmere, a little south of Rhinebeck. Crosby came from a family known for birds. His uncle was Eugene Schieffelin, the man we can thank for introducing European Starlings into North America. In 1890, he released one hundred Starlings into New York's Central Park as part of his (crazy) project to bring the birds mentioned in Shakespeare's work to the United States. His introduction of the Skylark and the Chaffinch on American soil was not a success; the Starling was too much of a success.

Though he was not an ornithologist, Crosby, who started birding as a child, kept meticulous records of his bird outings. Here is one example of his May census journal:

> Up at 3:20 standard time. Temperature 40 degrees, sky overcast. A Song Sparrow at 3:31, Whippoorwill and Night Heron at 3:37, White throat and Chippy at 3:39, Ovenbird at 3:54. Robin at 3:59; Barred Owl at 4:00, Catbird at 4:01, Wood thrush at 4:06, Phoebe 4:10, Grasshopper Sparrow going strong at 4:11, House Wren 4:14, Vesper Sparrow 4:15, Field Sparrow 4:16, Dove 4:20, house Sparrow 4:24; Chuker 4:27; Crow 4:28, then Oriole and Dove and Flicker at 4:35 (2nd one not till 5:03). 30 species noted by 4:48; 40 by 5:06; 50 by 6:06; 60 by 6:48—then breakfast. It drizzled at 6:03, rained at 6:50 and the sun was slightly discernible at 7:50. I left Grasmere at 8:30 and went to Cruger's Island arriving 9:25. Thence to Pine Plains, arriving at 11:30. I left there early in the afternoon, as the weather was vile, I had a very bad cold and was drenched and overbested.

Overbested. On some days even Crosby couldn't keep at it. Most days, though, he was up at three or four in the morning, clocking hours in the field. His records are so detailed I wonder how he had time to see the birds.

The Birds of Dutchess County, New York, which was first published a few years after Crosby's death in 1931 (from appendicitis), relied heavily on his records. The guide is a treasure of information about the birds that have passed through the county, as well as the stories of the people who searched for those birds. The biggest part of the guide is an annotated list of birds that gives a wealth of information about each species, including when and where it is most often found. The historical notes give a fuller narrative as to when a species has been seen in the county and by whom. A handy monthly graph indicates what months a bird might be around, and if the bird is transient or a permanent or summer resident.

One winter day I drove to the Franklin Delano Roosevelt Library in Hyde Park, forty-five minutes south of my home, to dig up Crosby's bird journals. Crosby knew FDR before he became president. Once the presidential library opened in 1941, Crosby's family donated his notebooks to the archives. The president's papers are housed in a windowless, sterile room. A few researchers sat at the spare desks, studiously poring over precious documents. We are forever fascinated by this man, I thought, as I watched their focus and wondered what hidden gems they had found.

The archivist gave me the long list of instructions that included no pens or food, and then rolled out a cart with several boxes containing Crosby's journals. I sat at my narrow desk, while across a divide, other researchers sat absorbed in their own reading, taking notes in pencil and snapping photos of key pages.

Crosby's bird notebook for 1909 is bound in red cardboard, worn smooth. Inside are wide ledgers like in a child's notebook, with blue-green lines on slightly yellowed paper. Reading Crosby's notes, written in beefy curves in black ink, I could practically see his pen as he made the birds come alive, winging out of the sky onto this page in front of me, telling avian stories from the past.

And then I started to weep, the tears pooling at the edge of my eyes, then slowly rolling down my cheeks. I wiped away these vagrant tears, with a quick swipe of the back of my hand, hoping the archivist didn't see the potential hazard to the paper that I had become. I puzzled over my tears. Maybe it was that the birds I read about were dead. Maunsell Crosby was dead. It all seemed so ephemeral. In that moment I understood lists in a new way. These lists are a record of the birds, and they are a record of life itself.

As I left the archives, passing in front of FDR's miniature white house of a country home, I felt euphoric. I pulled out my cell phone and called Becky, knowing that in her work as a historian she had often experienced this elation of the archives. She was happy to hear my giddiness, to share her own first moments of holding and reading bits of the past, how thrilling it was to find something that confirmed or challenged the ideas she was forming about a particular moment in time.

In this light-headed state where all thoughts seem bright and original, it became clear to me what was lost with eBird. I had given up the field notebooks and journals I kept when I first started birding. These small, colorful field notebooks line a row on a shelf in my home office. In them, I wrote down date and place, weather and what I saw, as well as whatever tips I got from other birders.

> Broad-winged Hawk. *They lead pretty secretive lives.*
> Kestrel. *All-star falcon. Hovering.*
> Winter Wren. *Most beautiful sound in the woods.*
> Green-winged Teal. *Fast fliers; small duck.*
> Golden-crowned Kinglet. *Dee dee deet.*

In rereading these notes, I could hear Peter's voice or that of another birder in the field. I could recreate the day, the weather, where we were, what we ate for lunch. From these notes I often wrote a full account, describing the outing in all of its complexity, if the birds were mating or skulking or feeding on fish, but also was I happy or had the sun vanished behind clouds. What we see or how we see it is shaped by longing, cold fingers, headaches, hurt feelings, or whimsical sunsets. These details etch the moment and the birds I saw more deeply into my memory.

In reading Crosby's journals, I had joined him as he rose long before dawn, breakfasted, endured the rain and exhaustion. Based on so little I liked this man, had a sense of his energy, his spirit. Once I read the story of his family, written by his granddaughter, I learned that he came from terrific wealth (which is too often the story in the birding world), that, like Peter, he lost his adored three-year-old son, that his wife left him, that he cared for his daughter, that he had a mistress in New York City. There's no sense of these wounds in his bird notes. The man I encountered in his notes was vibrant and attentive to the birds, even when he had a bad cold. What I imagined was that the birds went beyond being a solace; they became an affirmation of life, even a celebration. And I wondered how many birders had, through the birds, come to embrace their own lives.

Within weeks, Christina was hot on my trail, listing her birds through eBird. She turned all of her competitive urges that would have gone into track to birding; she wanted to win. Or, rather she wanted us to win, because whatever bird she saw, she alerted me. In turn, I drove her to whatever bird she had to see. In this way, on an almost daily basis, we swapped places, rotating from #1 to #3. She became the list mastermind, knew when I was number one or when we had lost that first place spot to Adrienne. Christina obsessively checked to see what we might have missed or just had to see; she knew how quickly the list refreshed, and what new bird Adrienne had found. Watching Christina I understood what real competition looked like. This included going out earlier and longer and later and daily, and driving great distances for one bird. Christina made jokes about sabotaging Adrienne's birding plans ("maybe I should let the air out of her tires," she smirked when she saw that Adrienne had pulled ahead). Christina's determination made me laugh, and I also saw that I still didn't have what it takes to compete. And yet, swept up in Christina's drive, when she texted saying "we have to go find that Greater White-fronted Goose," I got in my car and we drove down to an isolated farm to see it.

Adrienne was Christina's bête noire. In Christina's imagination, Adrienne was tall, imposing, and beatable. Once we finally met Adrienne we laughed because she is short, sweet, and unbeatable. Under her sweetness lay a determination to see every bird that passed through Dutchess County. After all of the nasty competition-fueled thoughts she had voiced about Adrienne, Christina felt silly: "I can't *like* Adrienne," she moaned, "but I do."

In the late 1920s, Maunsell Crosby invited Francis Lee Jaques, bird artist for the Museum of Natural History, and his wife, Florence Page Jaques, to join him on a Big Day in Dutchess County. A Big Day should really be called a Long Day, as it involves twenty-four hours of birding. One of the most famous Big Days is referred to as the World Series of Birding, held in the state of New Jersey, and the one that tallies the most birds covers all of

Texas. Peter did a county Big Day every year, driving around Ulster County like a lunatic and often finding an astonishing 140 plus species of birds. He then spent days after complaining of fatigue and recovering from the over-sugared state he always relied upon to make it through the day.

Peter, like most Big Day enthusiasts, raises money during his count, with people usually pledging a certain amount per bird. But that small philanthropic gesture didn't provide enough to outweigh the sheer excess of driving that is required for a Big Day (though some teams in the World Series move about on bikes), and above all, the explicit competitiveness of it. These two things combined means I've never been tempted into a Big Day.

Florence Page Jaques lets us into her Big Day in the wonderful first chapter of her third book, *Birds across the Sky*, published in 1942. At this start of her memoir, Jaques, new to birding through her husband, had no clue what a Big Day might entail. She describes her baptism as similar to "learning to swim by being thrown into Niagara Falls." The first shock was getting up at two in the morning. Once in the field, she felt "slightly out of place, like a cocker joining a pack of wolf-hounds." Bemused by the day, she dreamily comments that "this is a very odd occupation for three grown men." And yet she acknowledges the skill of the birders, describing Crosby as "one of the most gifted of field ornithologists. . . . His vision was keen as a hawk's; his hearing even more exceptional, both in its long range and its ability to distinguish between similar sounds."

Joining the Jaqueses and Crosby was Ludlow Griscom, who is often referred to as the "Dean of American Birding." Griscom, famous for his brusque sense of humor, is remembered for advocating for using field marks—distinctive characteristics of the bird like wing shape or leg color—that are diagnostic in making an identification. Since the use of field marks is now how most people bird, it's hard to believe there was a time before field marks.

Griscom was an original in many ways, was, in fact, such a character that Griscom-isms were well known throughout the birding world. They form a sort of found poem.

The Griscom-isms
Let's stop here and flap our ears

Check me on that one
    Well, we bumped that one off
        That's just a weed bird
            Now someone find a bird with some zip in it
        Just dribs and drabs left
    Please lower your voice to a howl
I don't like the look of that bird
    Put it down to sheer ignorance, incompetence and inexperience
        We got skunked on that one
            That's a 10 cent bird
                Having a good time?

Having a good time? I imagine that birding with Ludlow Griscom would be a good time as he had an encyclopedic memory, was a "genius in the field." But I felt sorry for poor Florence, thrown in with him on this, or in fact any other, Big Day.

The three men and Florence heard Barred Owls in the night and Whip-poor-wills in a pasture at dawn. They combed a "frog-haunted stream" for rails with no luck, and later in a "Botticelli colored field" she could "hardly refrain from out-singing the meadowlarks." Florence's contribution? A big black bird. "Simply a crow," her husband hissed at her.

At lunch, as Florence and her three men waited out a storm, Crosby explained to Florence how the data from this day would add to ornithological knowledge. "All these details may give us new ideas about bird behavior and distribution, or help in migration problems; perhaps in some manner which we don't yet realize." The "we did it for science" argument is a good one. But really on these long Big Days, everyone is there for the misery and fun of it.

At the end of the day, Florence Jaques was beat. She noted that the birds were no longer singing, just chirping. "And your laugh," Griscom observes in a detached way, "has become a titter." This comment, which seems in line with descriptions of Griscom, throws Jaques into a funk. Yet, when she learned that her group had seen 120 birds to the other party's 115, she abandoned her sulk. With this small but important victory, Jaques shaped this hard, long, hot day into a fun essay and declared that it was something she "might like to do again." Like so many birders before and after her, she was hooked. Her life with her husband and the birds stretched across the country, he illustrating them and she writing about them in eight nature books.

By the end of 2013, I lost my number one status in Dutchess County. Adrienne landed one bird ahead of me, with 184 species for the county. I didn't bother to call Becky to let her know.

Rusty Blackbird

Cruger Island Road, Dutchess County, New York

Dawn light filtered through early spring green leaves, warming my back as I walked down Cruger Island Road. *Tea-Cherr, tea-Cherr, tea-Cherr.* The Ovenbird's emphatic song filled the woods. The bird's brown feathers blend in with the leaf-strewn forest floor, making it a tricky one to see. I had to try, though, as it was my first Ovenbird of the year. As I scanned the dense hardwood forest, the smell of leaves decaying and the speckled light heightened my sense of deep woods even though I was minutes from the Bard College campus. The bird called again, loudly, as if it were right there.

I scanned low for movement, knowing that often the Ovenbird is near the ground. A cluster of bloodroot in bloom, little white flowers, as if birthday candles had been stuck in the earth, caught my eye. I lowered my binoculars to wonder at the fragile tenacity of these spring ephemerals. Just the word "ephemeral" made my heart clutch; *don't miss a minute of this*, I reminded myself. Then the bird sang again. So close.

I crouched, hoping that a new perspective might help my search. The bird sang. I cocked my head, listening to the hearty chips of a Song Sparrow, the distant call of a Red-bellied Woodpecker. The woods wrapped around me, still, expectant.

When I finally spied the bird it was sitting on a branch bowed like a stiff jump rope, six feet off the ground. It looked pleased with itself, as if it had found the perfect perch from which to broadcast its love. Its speckled breast puffed as it belted out one more song. Binoculars to my eyes, I could see the orange-ish racing stripe across its head and the white eye ring that made it look in a constant state of alarm. *The world is astonishing*, I agreed as I watched the bird hop down to snag an insect meal. It appeared nonchalant, shrugging off the heroics of its recent flight, from perhaps as far south as the Caribbean. It might be stopping in the Hudson Valley for a summer of eating and breeding, or it could be continuing north, into Canada. North or south, it lays its eggs in a Dutch oven–like nest, which is what gives the bird its un-warbler-like name of Ovenbird.

I loitered with the bird knowing that, so quick—maybe a day or two out—I would hear the Ovenbird song from deep in the woods and not stop to say hello to the bird. I'd be searching for the latest arrivals, maybe a Blackburnian or a Magnolia Warbler. This pull to the latest, the newest, the shiniest is easy to give in to. Some do it with cars and clothes, beer and restaurant choices. I've never been drawn to the next trend, happy to wear the clothes I had in high school or college (jeans and sweatshirts never go out of fashion), and I drive an old car. But the birds did bring out a greediness I gave in to; I wanted to greet the next new bird.

I was not moving into the next shiniest love. Rather, I was idle, like a painter in a Paris café, watching birds fly by, admiring the new outfits, the jaunty gait, the explicit sexiness of spring. I had settled into a contentedness, even happiness with my solo life, especially the freedom that came with that. I woke at five and chose to walk and find warblers. Or I followed an evening urge to paddle the North Tivoli Bay to greet a Green Heron. In my time birding with Peter, I had paddled less and less, finding it easier to bird with both feet on solid ground. But I had missed my kayak and the water world I knew so well. The Hudson River and the Tivoli Bays welcomed me back, and I took on the challenge of joining that water world with the bird world. The two were a natural combination, my kayaking taking me to Least Bitterns or American Bitterns, birds I would not easily see from shore.

This freedom to do what I wanted felt a luxury. I didn't think of this as *getting to know who I am*, that cliché of the single. I was getting to know the birds. But it turned out that as I learned more and saw more, I also saw myself more clearly. The qualities that I curated day after day in looking for birds—patience, attentiveness, a loving devotion—had leaked into other aspects of my life. I saw it most when I found myself more patient as I took care of the elders in our small family, my father's cousin and my uncle. I sat through long lunches or for hours bedside; I listened to the same stories over and over, and laughed or smiled as if it were the first time. And I never tired of talking about the weather, the one sure subject beyond doctor visits.

Maybe I had, like the birds, molted. The birds do it every year shedding one set of feathers and growing into another. During that period of molting they are often unable to fly, leaving them vulnerable. Through the winter months I, too, had passed through a phase of feeling exposed in my loneliness. Now, in the spring light, I had donned my new feathers. I moved through the world with lightness, returned phone calls to friends, awoke astonished that, like the Ovenbird, I was pretty pleased with my perch, my perspective on the world.

On this walk down Cruger Island Road, a Downy Woodpecker whinnied its flight, and a White-breasted Nuthatch quanked in its plaintive voice. In the distance a woodpecker rapped against a tree as if sending a message in morse code. A Pileated? A Great Blue Heron flew over, its long legs dangling behind, its wings huge and awkward, more dinosaur than bird. At the bottom of the hill, a Song Sparrow delivered its charge: one, two, three! While a Red-eyed Vireo commenced its irksome loop: up here, in the tree, here I am. A Catbird in a nearby bush talked to itself or criticized me, I couldn't tell, and a Yellow Warbler sang its sweetness for itself alone.

Orienting myself through song felt like the great gift of birding. Up to now, my greatest tap into memory had been smell: of a lawnmower and wet grass taking me back to summers spent at my grandparents' house at the Indiana Dunes. Or chestnut trees in bloom reminding me of midsummer in the southwest of France. After a few years of listening to birds I realized how songs brought forth memories, added to the texture of my past. Now that I could name the birds, I knew it was the Gila Woodpecker that laughed in the Tucson mornings; the Northern Mockingbird and the Red-eyed Vireo signaled home in Pennsylvania; the Herring Gull screamed of summer and family on Cape Cod; the Veery—my spark bird—that spiraled its song of silver from the woods as I walked out at dusk after a day of rock climbing at the Gunks. And the Ovenbird would always remind me of deep woods hikes. Learning bird songs had woven another thread of memory into my landscapes.

At the end of the gravel road, a causeway that bisects the North and South Tivoli Bays leads out to Cruger Island. I know the causeway in fall when the leaves drop and the phragmites of the North Bay turns strawlike, swishing in the wind. I know it in winter when I track out through the snow, searching for footprints not human, both Bays frozen, smoothed by wind that races off of the Hudson River. I find it irresistible in spring, when it is lush with wisteria and honeysuckle, as if I were in a southern bayou, and possibility lurks in every bush.

Marshes are special spots, ones that most people ignore or even fear, the ground soft, the air abuzz with mosquitoes, the mood suggestive of ghosts. I am thinking of the opening of Charles Dickens's *Great Expectations*, the first adult novel I fell in love with. "Ours was the marsh country, down by the river, within, as the river wound, twenty miles of the sea." Our hero Pip, just as big as his name implies, is there on a "raw afternoon," in this place overgrown by nettles. The overall sense of doom is cast as Pip peers at the names of his parents and siblings on a tombstone in the cemetery. Pip is a "small bundle of shivers" when he is caught by a fearful man, dressed in "coarse grey." The man is an escaped convict. He picks Pip up and shakes him upside down to extract a bit of bread from Pip's pockets. It's not hard to read this passage and understand that this convict will, in all ways, turn Pip's life upside down. Aged twelve, though, I was not conducting a literary analysis. I was shivering along with Pip, the encounter with the convict as thrilling as the land it was set in.

Convicts and their secrets lurk in marshes. So, too, do secretive birds: Virginia Rail, sliding between the reeds, Least Bittern, blending in with the grasses, and the Sora, the bird I most wanted to hear—such a crazy whinny of a song—and see: such a delightful, compact body with goofy large feet. I listened, hoping to hear the bass beat, *glump dee dump*, of the American Bittern emerge from the cattails. The stealthy birds of the marsh are joined by the extroverts: kingfishers cackling as they fly on a special mission, and the Red-winged Blackbird insisting on its *chuck*, or *peer*, or *ok-la-dee*. Marsh music is that of a jazz band, odd angled, with unexpected riffs and elegant solos.

I did not meet a convict in the marsh, but rather Christina, taking a quick break from working on her senior project. I wasn't surprised to see her—she was making the Bays her home, walking out daily to check on the bird life. And I was glad to see her there, crouched in the muddy causeway, peering into the reeds. I walked up slowly.

"There's a Least Bittern in there," she whispered.

I crouched next to her and played a game of Where's Waldo, the lines of the reeds a perfect match for the lines down the bittern's chest. I let my eyes soft focus, and there I saw texture that varied, the soft chest of the bird as it stood so impossibly still.

"Nice," I breathed. "Did you hear the Ovenbird?"

She nodded.

Christina had to run back to campus, so we said our goodbyes. I continued on, sloshing along the muddy causeway, which is just the width of a car, or these days a DEC-driven ATV, the rangers checking on the Bays. To the north spread a cattail-and-phragmites-laced waterland, the channels deep. That was where I kayaked most often. To the south, past a trough where snapping turtles loaf and from time to time a northern water snake sallies by, rests the open, shallow South Tivoli Bay, where ducks skid in to dabble. That was where Christina and I had found the Canvasbacks.

At high tide, the water rises and drowns the causeway, making it impassible except in muck boots or waders. I wondered what John Cruger, who built his house on the island, did in the mid-nineteenth century. Perhaps the causeway was above ground then as the flooding of recent years is mostly thanks to beavers. They were on the brink of extinction when Cruger lived in his island house. Now protected, beaver have returned to control the water flow and sculpt the landscape. Not everyone appreciates beaver architecture, how roads and backyards flood. Out in the Bays, though, no one pays much attention or cares about the industrious rodents. They are left in peace as they build new stick domes, forcing the water to move in unexpected meanders through the marsh.

A Bald Eagle flew over, its shadow blocking the sun for a moment. This eagle had the white tail feathers and head of a mature bird. It is a bird I take for granted, though I know I should not, as eagles are a species that toed the line of extinction. There was one central thing we had to do to save this big bird in decline: eliminate DDT from the environment. It was Rachel Carson in writing *Silent Spring* who alerted the world to the dangers of DDT. For eagles, and other larger birds like the Osprey, the DDT makes for thinner eggshells. The simple natural course of nesting can crush the fragile eggs. Carson's work was central to legislation that banned DDT. Slowly the chemical has almost disappeared from the ecosystem. That accomplished, this magnificent bird is back and for years had a nest on Cruger Island. It was one of over a dozen nests in the Hudson Valley, when in the 1970s there were none.

The air around me filled with whines, cackles, and whistles, what Henry David Thoreau called the salt and pepper of the air. The Red-winged Blackbirds swooped over the cattails and grasses of the Bays like street boys on bikes, dodging traffic, jumping the curb onto the sidewalk, and just barely missing the stop sign that means nothing to them. Put another way, the Blackbirds "twist, tumble, tangle," glide and curvet, which rhymes with epaulet, those dashes of scarlet on the shoulders that make the Red-winged Blackbird so handsome. These are the words of the poet Robert Penn Warren in his poem, "Redwing Blackbird." It is not the excess of life Warren experienced seeing the birds, as I did there, swimming in the sounds that promised the future, that spoke of abundance of flowers, birds, life. The poet tilted toward the passage of time, toward death:

The globe grinds on, proceeds with the business of Aprils and men.
Next year will redwings see me, or I them, again then?

Is the poet anticipating his own death or that of the birds? Or perhaps both? Was it possible to imagine the death of all of those Red-winged Blackbirds? Probably not, as the abundance, the sheer force of them appeared unstoppable. No doubt, though, people had the same thoughts about the ever-present and plentiful Passenger Pigeon, the bird Alexander Wilson saw two billion of in the late 1700s, that was now extinct.

It was fearing the end of a species that had, in part, led me down Cruger Island Road on this early spring morning. I was hoping for a cousin of the Red-winged Blackbird, a bird I was afraid would not appear. The Rusty Blackbird.

The Rusty Blackbird is not a bird that asks for love. Though it might be named for the rusty color that in fall adorns the neck of this jet-black bird, its name might describe its squeaky, almost unpleasant song that hardly passes for a song. "Their united voices make a volume of sound that is analogous to a bundle of slivers. Sputtering, splintering, rasping, rending, their notes chafe and excite the ear. They suggest thorns and briers of sound," wrote John Burroughs. Then adds: "and yet are most welcome." Indeed.

Two years ago, Peter had introduced me to the Rusty on the muddy Cruger causeway. "Look now," he said, "this is a bird that will be extinct in our lifetime." The bird appeared ordinary enough, a black bird, except for the eyes, which are a spooky white. At that time I doubted Peter's pronouncement; it seemed too filled with doom. Then I came to learn that he might be right: Rusty populations have plummeted by 88 percent in the past century. Scientists estimate that somewhere between 158,000 and two million remain.

Though I accept these numbers—estimating bird populations is complex—to me, the difference between 158,000 and two million is vast. Nowhere else in my life would I accept such a range in numbers. (If a contractor said, I'm going to build you a house, it might cost $158,000 or it might cost $2,000,000, I would not sign on!) Yet that is as accurate as it gets. In 2005 scientists formed an International Rusty Blackbird Working Group, and we still know little about this bird. Part of it is that the Rusty breeds in inaccessible boreal bogs. Tagging, then refinding those tagged birds, has proven near impossible. A 2010 article in *The Condor* is titled "Rusty Blackbird: Mysteries of a Species in Decline." I can like a science article that admits to the mystery. Is it not all a mystery?

The actual mystery that these scientists are mulling over is why the decline is occurring. There is no one simple answer like DDT with the Bald Eagles. There's habitat destruction, of course, the given with any steep drop in population. The woods the Rusty needs on their wintering grounds in the southeastern United States have been transformed into agricultural land. The boreal wetlands where they breed have been lost to peat production, timber harvests, oil and gas exploration. Added to this scenario is mercury, which is naturally occurring, but in acidic wetlands, where the bird breeds, it converts to its toxic form, methylmercury. This likely kills the embryos of the Rusty. In other words, there are a lot of challenges for the Rusty. And the question is, which ones of these can we change, and which one is most urgent to fix.

The article on the Rusty ends with a bit of philosophic puzzling: "the most important looming question is do we have the will and the resources to address the mystery of the Rusty Blackbird's decline?" Gaining public sympathy, which in turn will move politicians or granting agencies with dollars to get the work done, is easier to do with charismatic birds like the Whooping Crane. The Rusty is not a sexy bird. It's not even a bird known to most people. So I doubted people were rushing to their wallets to help out. Yet the Working Group is testimony that some have the will to organize, research, understand, even fight to save. Some do not want to fail this rusty black bird.

In 2014 the Working Group instituted the Rusty Blackbird Blitz, asking birders across the country to focus on the Rusty during migration then log those sightings into eBird. I took to it like a convert, imagining sightings would somehow help, each bird seen and logged in a vote for hope.

Cruger Island Road where the Rusty stops on migration is one place where Maunsell Crosby spent hours conducting his meticulous counts in the early part of the twentieth century. On April 19, 1925, there were several hundred Black Ducks, Mallards, and American Wigeons, which he calls Baldpates. On October 7, 1927, Crosby rose at 4:10 in his house outside of Rhinebeck and by 5:30 arrived at Cruger with Ludlow Griscom. They heard a Great Horned Owl; Griscom recorded a Gadwall flying over, and then in came a flock of American Scoters with one Surf Scoter tucked in there. There were Pintail and Green-winged Teal, a Greater Yellowlegs, and Fish Crow. Though Cruger is still full of birds in the twenty-first century, it's nothing like what Crosby describes.

It was Crosby's notation from much earlier, on April 7, 1909, that caught my eye. There he recorded his first Rusty Blackbird for the year. It's the nineteenth species he found in 1909. Looking at Crosby's bird lists from the past made me wonder: What birds will fly, sing, breed here in ten years? Twenty? One hundred? Would the Rusty somehow, by mystery or legislation, by the work of determined conservationists or by their own will to live, make it?

I didn't linger long with these unanswerable questions, my soul unwilling to bend toward loss. I preferred to continue searching for the birds, wanted to celebrate all of the new life flying, eating, even nesting about me. The point is to carry on. But I did not want to carry on in silence. Rather, I wanted to talk of, and write about, the Rusty, the Henslow's Sparrow, the need to save swamps and fields, the need to stop the population plummets of so many birds. I wanted to take students out and show them the birds, have them, like Christina, see the birds and feel the importance of saving them.

I saw the path that many people took, devoting their lives to the birds, to the environment. I admired, even envied them. But I sensed I was too late for that, or maybe that I could not live with that daily sadness, anger, frustration. I would do what I could do, teach and write; I'd be a part-time crusader, as Ed Abbey called himself. Abbey would have counseled: *It is not enough to fight for the birds; it is even more important to enjoy them. While you can. While they are still here.*

I moved onward, flushing two ducks. A flash of cinnamon, white, and black; the ducks whinnied like irritated babies. Wood Ducks. I felt sorry that I had put them up. My feet squished through the mud where a black, shiny leech extended, in hopes of latching its next meal. A Catbird played a medley from low in a bush, and a flock of Common Grackle gossiped from a bare tree. As rich as this was, I did not hear what I hoped for. One ornithologist described the Rusty Blackbird song as an "unoiled windmill." Ornithologist Edward Forbush wrote that the *chuck* call of the Rusty can be mistaken for a frog, but that the song is unmistakable: "Their chorus then is a mixture of chuckling and shrill squeals or whistles unlike any other sound in nature."

I stopped from time to time, straining to hear the chuckle of the Rusty. Goldfinch. Robin. Kingfisher. Fine birds, but not my Rusty. After half a mile of flat walking, due west with the Catskills in my view, I reached the end of the causeway. There, gravel shores up the steel rails of train tracks that have, since 1851, followed the shoreline north and south along the river. I stepped up near the rails and scanned toward the North Tivoli Bay, hoping to see a Northern Harrier gliding over the brown grasses, now flecked with new green.

On my return, I continued to listen. Nuthatch. Cardinal. Common Yellowthroat. Marsh Wren. No Rusty.

What would it mean if the last Rusty left this planet? Theodore Roosevelt, then governor of New York, said it best in a letter to Frank Chapman written on March 22, 1899:

> My dear Mr. Chapman . . . I need hardly say how heartily I sympathize with the purposes of the Audubon Society. I would like to see all harmless wild things, but especially all birds, protected in every way. . . . Spring would not be spring without bird song. . . . When the Bluebirds were so nearly destroyed by the severe winter a few seasons ago, the loss was like the loss of an old friend, or at least like the burning down of a familiar and dearly loved house. . . . The destruction of the Wild Pigeon [Passenger Pigeon] and the Carolina Paroquet has meant a loss as severe as if the Catskills or the Palisades were taken away. When I hear of the destruction of a species I feel just as if the words of some great writer had perished; as if we had lost all instead of only part of Polybius or Livy.

Of course Polybius or Livy are a bit lost, read by only the most fervent classics major, which is a species of student perhaps as rare as the Rusty. Still, I appreciated Roosevelt's scale, his sense of urgency. I faced west, toward the Catskills, imagining the undulating land there suddenly flat, the mountains gone. That might happen, but in a time frame in the millions of years, not in decades. And I couldn't imagine so far into the future, I couldn't picture the landscape there flat, the mountains so much a part of the shape of the earth. As I walked home, that was the image I wrestled with; the world without the Rusty was not a world I wanted to live in.

Dawn Chorus

Thompson Pond, Pine Plains, New York

On May 10, 1942, President Franklin Delano Roosevelt rose at three in the morning at his home in Hyde Park. At four he left his sleeping guests, the crown prince and princess of Norway, and drove off with an esteemed group that included Allen Frost, who was considered by many to be the bird authority of Dutchess County; Ludlow Griscom, one of the best birders of his generation and the man who led Florence Page Jaques into her first Big Day; and FDR's distant cousin, Margaret Suckley, known as Daisy. Daisy Suckley, probably most famous for giving FDR his Scottish Terrier Fala, was a long-time member of the Rhinebeck Bird Club, a tony group that included Astors and Vanderbilts. Daisy knew her birds.

As Griscom settled into the car, FDR told him, "We call this car the Queen Mary." The car had bulletproof doors; hand grenades rested on the floor. A long, black Secret Service vehicle closely followed the presidential car, while a third car trailed further behind, driven by the curator of the Roosevelt Library, James Whitehead, accompanied by Poughkeepsie lawyer Raymond Guernsey. The group was pointed toward Thompson Pond on a bird-watching venture that they had been planning since March. "I wrote him about going with Frost to hear the 'Bird Chorus' at Thompson's Pond on May 10th!" wrote Daisy Suckley in her diary on March 21. "He is *thrilled*." Daisy, who dedicated her life to adoring and serving her distant cousin was overly fond of the exclamation point. In this case, it seems about right.

The Secret Service men were in a tizzy as the car lazied in the dark, east and north, through the back roads of Dutchess County, the top open in the big car so as to hear the bird calls. What they heard included *many* Whip-poor-wills, a bird I have yet to hear or see in Dutchess County, though early accounts of spring birding almost always include the Whip-poor-will's song. It's a song you can't miss, because the bird really does call *whip-poor-will* on an endless loop. Across the Hudson River in Ulster County in the nineteenth century, John Burroughs reported once hearing a Whip-poor-will sing 1,088 times. I marvel that Burroughs sat or stood and listened, counting to such a number. How long did that take? And what made the bird stop at 1,088—had it finished what it had to say? The bird's endless song and the man's devoted counting, what here is more wondrous?

The Roosevelt party arrived at Thompson Pond, just south of Pine Plains, under a "damp and cloudy" sky. Whitehead, in a short article published in *The Conservationist* (a publication of New York State's Department of Environmental Conservation), described the morning. The president spoke in a half whisper, in awe like all of them as each bird chimed in to create a symphony of sound. They heard Virginia Rail grunting their *kiddick kiddick kiddick* like two flat stones knocked together, Sora making an otherworldly whinny, American Bittern galumphing from the near dark, Marsh Wrens bickering from the reeds, and Pied-billed Grebes intoning *cow-cow-cow-cow*.

I knew Thompson Pond and the wonder of birds there, but nothing on this scale as reported in Whitehead's article. Thompson Pond was then and still is a near perfect habitat for ducks and marsh birds: it's a wide, shallow body of water, a glacial kettle pond that flows south into Wappinger Creek, which in turn makes its way to the Hudson River. The slow-moving water is flanked by cattail and marsh grasses; at the southern end rests a bog. A trail about two and a half miles long loops around the pond, offering views out, where, depending on the season, I can see Canada Geese, a range of ducks, and almost always a Great Blue Heron tiptoeing the shallow borders focused on its next meal. Stissing Mountain, at 1,403 feet, shadows the western edge of the pond. Raptors float there, catching the wind currents along the ridge. Golden Eagle have been known to winter on the mountain. It's a special enough spot that in 1958 the Nature Conservancy purchased the seventy-five acres for preservation. In 1973 the U.S. Department of the Interior designated it a national natural landmark.

The presidential party parked at the northern end of the pond, where a marshy blend of snags, cattails, and phragmites straddles the narrow strip of road. When the Secret Service brought out a big searchlight, a hush descended. No bird wants to sing under a spotlight. Griscom asked Roosevelt to turn off the lights; Roosevelt responded: "There is one person that the President of the United States can not tell what to do, and that is a Secret Service man. If you want to go back and plead with them, go ahead." Griscom, with his élan, pleaded or asked or demanded. They turned off the lights.

Ludlow Griscom had initially hesitated in joining this outing. He considered FDR's "shenanigans" outrageous and had voted for Wendell Willkie in 1940. His wife Edith argued with him: "Ludlow, this man happens to be the President of the United States, there's a war on, a very serious war, and he is absolutely exhausted. He's back there at Hyde Park resting. It's your patriotic duty to go and do anything that can relax his brain in any way, shape or form." This might be the only time someone has birded as a patriotic duty—but what a great duty it was, to relax the president's brain.

Relax the brain. What an odd phrase, as if the brain were a muscle. It's not, but our minds do need to rest, and not just in sleep or with a glass of wine. Generations of presidents have looked for ways to escape the endless flow of responsibility, whether through fly fishing or golfing. FDR escaped with the birds.

I doubt FDR and Griscom talked politics, leaving differences behind in favor of enjoying the birds. Imagine a world where Congress sauntered out to bird together, Republicans, Democrats, and Independents shoulder to shoulder working out the identification of the Mourning Warbler. I'm sure it would be a more genial, a more productive, governing body.

Politics aside, Griscom was the man you wanted on any birding outing. No doubt, on this May 10 count, Griscom was the one who spotted the Black Tern and located the twenty-three species of warbler, including the Cerulean, a bird that has always been rare in Dutchess County, and Golden-winged, a species that was common enough but now rarely seen (and never seen by me). I wondered what the president added to the list.

In the opening of his article in *The Conservationist*, Whitehead tells a story of FDR's "selective" eyesight. "Once, when somebody complained to Eleanor Roosevelt that FDR had not spoken to him when he passed nearby, Mrs. Roosevelt replied that the President had blamed such incidents on the fact that he was nearsighted—'but,' she said, 'that has always seemed strange to me, for, as long as I have known him, he could always identify any bird he happened to see.'" Which is another way of saying: we see what we want to see. Often what we want to see is what we love.

If the president did not add birds to the count, what he surely added was entertainment. He told stories. One involved army patrol planes that reported twenty-one German submarines near Bermuda. The United States had been in the war for five months at this point. A state of nervous anticipation took hold as navy bombers mobilized. The bombers flew out only to count twenty-one whales lolling in the warm waters. The president laughed at his own tale of army ineptitude. Another story involved Steve Early, his White House press secretary. The president requested he shoot some ducks, tasty ducks (though the species is not given). This makes me wonder when last a White House secretary hunted on presidential command. A game warden stopped Early as he returned from the field, dead ducks in hand. The ducks had been shot out of season. Early pleaded that they were for the president; the warden insisted the $25 fine per duck be paid. I like this ethical game warden who did not let even the president slide past the laws.

It is fun imagining sitting under a misty sky at Thompson Pond listening to the president. But sitting about and chatting is not an efficient way to count birds. I wonder if Griscom was enjoying himself or eager to hustle off to Cruger Island to see what else they might catch in the early morning hours.

A black-and-white photo of the birding party, taken by Suckley, shows Roosevelt seated in the car, flanked by the other birders. He appears relaxed after his morning listening to birds, talking about birds, and telling stories.

Suckley in her account of the day wrote: "He seemed to really enjoy every minute. [Birding] is the kind of thing he has probably given up any idea of ever doing again, so it did him lots of good." That he didn't have time for birding doesn't surprise me. I found that my birding occupied a generous part of my day, every day, and even more so during spring migration. This might explain why there are few presidents who paid attention to birds. Those presidents who cared for the birds include Thomas Jefferson, and Theodore Roosevelt, who was attentive to all things in the natural world and kept a bird list from the White House. His interest in birds began early in life. While in college, he published a pamphlet titled *Summer Birds of the Adirondacks*. He, more than any other president, worked to protect the natural world and birds, creating five national parks and fifty-one bird reserves. Other than the Roosevelts, Jimmy Carter (whose bird activities occurred mostly after he held office) was the only president to bird.

Though I sometimes felt awkward about the hours I spent with the birds, I never felt that they took away from my work. Rather, the birds and my time in the field fortified me for what I had to do, and the energy and my *relaxed brain* made me more efficient.

"In that far-off silent place, with myriads of birds waking up, it was quite impossible to think much of the horrors of war," writes Suckley in her diary. This, then, is what birds can do even for the president of the United States: allow him a few short hours to forget and to fortify him for the work he had to do.

In the mid-1980s, I joined a friend who had arranged to have tea with Margaret Suckley at her fairy tale of a house, Wilderstein, on the outskirts of Rhinebeck. In her nineties, she lived there with help, while the house, perched on a hill with inspiring views of the Hudson River, creaked around her. We sat in the kitchen, the one room where the ceiling appeared without cracks. She was sweet and proper, and the tea was lukewarm. Though I don't remember the specifics of our conversation, I liked her enormously, her frail properness, her sense of humor. Mostly, though,

I was just so pleased to be speaking with someone who had lived through such a big moment in history and had had a front seat at the arm of this formidable president. What I didn't know then was how intimately she knew him.

After Suckley died, friends found a suitcase with her journal as well as letters to and from FDR. They reveal her relationship with her distant cousin, one that involved tenderness and generosity on his part and adoration on hers. Whether they were lovers isn't clear—she's circumspect in how she describes those afternoons alone with him eating buttered toast and drinking tea, those rides through the Dutchess County countryside, or down Cruger Island Road to watch birds. In reading about their outings I realized that birds gave FDR this: time with one of the women he loved.

FDR loved several women, and through those affairs the love for birds was a constant. That deep love had been with him since his childhood. The presidential archives hold his bird journal from 1896. It's a pocket diary, the kind with the swirly ink on the outside. On Wednesday, January 1, 1896, the weather was fine with a wind and the thermometer at twenty-five degrees. On that day the fourteen-year-old Franklin saw chickadees at 9:00 a.m. and juncos at 12:00 noon. On January 5 it was cold and clear, and he saw crows, nuthatches, chickadees in the morning and heard jays. Monday the 6th was bitterly cold, negative fourteen, but in the p.m. there are crows and at 9:00 a.m. nuthatches.

Through the weeks, Roosevelt demonstrated great patience as he kept at his sightings. In this daily way, he added other species: Downy Woodpecker and Brown Creeper on Friday, January 10, and goldfinches on January 21. On January 23 he saw a flock of Pine Grosbeaks eating the seeds from the pinecones at the top of some tall conifers; the young Roosevelt noted that they were all in gray plumage, except one, where sunlight highlighted the rusty red. Though Roosevelt did not like shooting the birds—as some birders did at that time—he shot at these grosbeaks, missing his mark, though he was able to collect some feathers.

That winter of 1896 grosbeaks settled on the Hyde Park property, fifty seen on the fourteenth of February. He describes these robin-sized birds as "very tame and unsuspicious and I have been within 15 feet of them several times." He often located them by ear, which might or might not be a challenge. *Sibley's* offers this as the voice of a Pine Grosbeak: "relatively low, lazy, unaccented warble of soft, whistled notes." Edward Forbush writes that a Pine Grosbeak sounds "like a lost chicken," a "peculiar, querulous, whistled *caree.*" These sound like such unrelated descriptions, two different birds. When I listen to a tape of the Grosbeak what I hear is rich, a warble indeed, but not a bird that is at all lost.

On January 29 Roosevelt shot one Pine Grosbeak. Roosevelt was lucky to have these birds visit as Pine Grosbeaks are a species that live in northern climates, often at high elevations. Since 1993 none have been seen in Dutchess County. He sent the bird off to be stuffed and mounted.

On February 18 Roosevelt traveled to New York City where he met Frank Chapman, then an assistant at the American Museum of Natural History. Imagine the fourteen-year-old Roosevelt knocking on the door with his birds in a bag, eager to meet this star in the ornithological world. "I am to send about 1 dozen Grosbeaks to the Museum for Local Collections," he wrote in his journal. He did send Chapman several specimens, which are still part of the collection.

As with many journals, FDR's notes diminish as the months unfold. From Thursday, June 11 until August 6 there are no entries. During that time, he was off to Campobello Island, in New Brunswick, Canada, where the family summered. On August 6 he was in Germany spotting Green Woodpecker, Great Spotted Woodpecker, Lesser Spotted Woodpecker, and European Goldfinches. Come fall there are no entries, as he attended Groton for his first year of schooling away from home.

I wish there were more entries. Reading FDR's bird journal makes this formidable president seem like you or me, a kid with curiosity and a stick-to-itiveness that any birder recognizes. One of FDR's biographers claimed that he enjoyed informality. He addressed everyone by his or her first name, including the crown prince and princess of Norway he left sleeping at his house that May 10, 1942, morning. In his many fireside chats, he wanted to give people a sense that he was speaking to them individually. Reading Roosevelt's bird journals, I felt like I was at his side, seeing those birds with him.

I had long had a sense of this intimacy with FDR, not just because his house in Hyde Park sits a half hour drive from my own, our geographies shared, but because my father grew up with FDR as his president. When my father was five, FDR was elected president. For the next twelve years, until he was seventeen, he knew only one president, one who was a part of the fabric of everyday life, one admired and respected and loved by his family and much of the country. My father spoke of the day in 1945 when FDR died as if losing a family member, the country in tears. I never had that sense of affection for a president. When I was eleven the TV never turned off as the adults watched the Watergate trials. I grew up thinking presidents were crooks.

This story of FDR, Griscom, Crosby, and Suckley eating sandwiches for breakfast and hearing Sora gave me another level of intimacy with the president and brightened my imagination whenever I birded Thompson Pond. I relished the texture and depth that FDR's Thompson Pond outing gave to the land, to the birds. This was different than the history of birders I'd savored as I walked Snake Bight Trail in the Everglades or when I read about Florence Merriam Bailey making lunch in the desert of Arizona where I birded with Deb. This morning outing and the people involved created a bridge between the larger world and the insulated one of birders. Birding wasn't separate from life; it was part of a larger whole.

And so, inspired by my romance with this bit of presidential birding history, I decided to undertake a modern reenactment. I wanted to relive FDR's day, to listen to the dawn chorus, and compare his bird world to the one I experienced. In venturing out in this way, I felt like I was merging with history.

On May 10, 2013, I woke at three in the morning. The crown prince and princess of Norway were not sleeping in my house, just my two cats, (Princess) Clementine and (Mr.) Slim. Stepping out of my house, a cold silence greeted me. In the dark all dreads are amplified; so too are all hopes. At my doorstep, I strained to hear my first bird. Where was my local Great Horned? Barred? Shouldn't something be rumpling about? The un-sound of the night unnerved me.

I swung by the Bard campus to pick up Christina. She stood in the dark, no one else on campus awake. I handed her a granola bar, which she happily accepted. I remembered those college days when I was always hungry.

We drove the back roads of Dutchess County in silence, the car windows down so we might hear all of those Whip-poor-wills that the 1942 crew had reported. Christina leaned with her head out the window, taking the cold air on her face like a Golden Retriever puppy. The temperature had dropped to thirty-three degrees so I turned up the heat, cranked the seat heater. Surely the president had warmer weather to greet him as he drove out for the dawn chorus. Though our cold ears strained for any peep or whip, all we heard was the rough whir of tires on asphalt.

After a half hour of navigating narrow roads, we arrived at a pullout near the trailhead at Thompson Pond. The silence enveloped us as we whispered laughingly about the cold. There was no reason to whisper except that I sensed we were in a church, a sacred place where loud talk would be scolded by the Red-winged Blackbirds. A Barred Owl let out its calming, familiar hoot, *who cooks for you?* And we both did a little skip of excitement as we walked down the road that runs where FDR and his companions had installed their cars. Soon some Canada Geese started to sharpen their honks from the reeds. A slow, simple trill announced a Swamp Sparrow.

The sky shifted from black to gray, a thin stripe on the horizon growing wider as we stomped (my feet wishing for warmer socks). We had arrived before the dawn chorus, just as the birds were tuning up. First, the whine of a Catbird, then the *conkla-dee* of a Red-winged Blackbird. Then Tree Swallows twittering as they ricocheted over the water. So fast, more birds joined in, the sounds louder, more insistent, more joyous.

"Wait, what is that?" I was on high alert for a Virginia Rail and desperate to hear a Sora, both birds on the list created by FDR's party, and good birds by Dutchess County standards.

The Sora is three ounces of bird, slipping through the reeds on its yellow-green legs. It is surely not the most beautiful bird with its gray chest and stout yellow bill. What enchants me about the Sora is its song, a sharp, clear whinny that descends the scale in a manic, playful way. It electrifies any marsh it lives in. And I never hear it enough.

Edward Forbush recounts in his guidebook the story of a man who on one day in 1880 shot 181 Sora on the Housatonic; on a typical outing a hunter could easily bag a hundred. Though less plentiful, Sora are still hunted in thirty-one states, with liberal bag limits. In New York you can take eight a day. I find this horrifying. Unacceptable. Though the Sora is the most abundant rail in North America, populations are not known. What we do know is that the wetlands they inhabit are threatened. I strained with no luck to hear even one of these birds that FDR heard in 1942.

We ticked off each bird as they joined the growing chorus: the grating *bubble-zee* of the Brown-headed Cowbird, the familiar yodel of the

American Robin as it hopped down the road, the *witchity witchity* of the skulking Common Yellowthroat. The early songsters were doing well, though to my ear they could have used a finer conductor, the several songs not yet coming together. Still, I was savoring this free concert. Black-capped Chickadee. Wood Thrush. Eastern Phoebe. As we stood on the road, we could watch as the sun smeared the horizon, a sliver of light. Veery. Ovenbird. Louisiana Waterthrush.

"Blueberry Gnatcatcher!" Christina called, and I laughed. Her young ears always heard the Blue-gray Gnatcatchers before me. We walked down the road toward the woods to see what warblers we might hear. Yellow-rumped. Black-and-white. Redstart.

"Breepers!" The Great Crested Flycatcher sat at the top of the tree weeping and breeping.

We entered the path that threaded through the woods, heading south around the pond, something that FDR and his group did not do. At a near overlook we were surprised to find a man leaning against a tree, binoculars snug at his chest, communing with the dawn.

"Have you heard anything special?" he asked.

"Yes," I said, because to me they were all special. "Actually, just the standard operating birds," I added, not wanting to mislead him.

"I'm not a birder. Can you tell me what birds are standard?"

I puzzled aloud over his binoculars.

"Oh, FDR came here on May 10, 1942," he explained.

After sharing quirky bits of FDR history, which Christina ignored as she scanned the water looking for ducks, we continued on, the woods close.

"Is this a thing?" Christina asked.

"I guess so," I said, realizing that maybe this was the first ever meeting of the May 10th-ists.

The sun had not yet started its job, so we moved quickly to generate our own warmth. The woodpeckers—Red-bellied and Downy—gave us a rhythm to walk to, and then the Warbling Vireo started its beautiful, jumbled song. This ordinary-looking brownish bird sang and sang and sang as we walked the two and a half miles around the pond, balancing on wooden planks over the wet trail. We heard or saw twenty in all. FDR and his group also heard the Warbling Vireo, one of thirty-nine species of birds we had in common. I'm jealous they had twenty-three warblers to our ten, while I wonder what on our list would have intrigued Ludlow Griscom. Perhaps our Pileated Woodpecker or the *many* Cardinals, a bird they did not see. Or maybe he would have been impressed with the five Red-bellied Woodpeckers. Both the Cardinal and the Red-bellied Woodpecker have expanded their range north, becoming common locally in summer or winter. In 1942, they would have been unusual sightings.

From 6:30 to 7:30 a.m. it rained on the presidential party. At seven, they all ate cheese sandwiches and drank coffee, then FDR returned to Hyde Park with his Secret Service detail, to get on with the business of running the country. At ten the weather cleared as the rest of the party continued on to Cruger Island, Mt. Rutson (known to birders now as Ferncliff, a great spot for warblers), Vanderburg Cove, Poughkeepsie to find Purple Martins nesting at the ferry, and Waltons (not a place I know of to bird), finishing the day at 7:20 p.m. with 108 species, including some special birds like a Henslow's Sparrow.

The checklist for the day is pocket-sized, yellowed, opening like a little book. Daisy Suckley sent a note asking, "Would the P. be so kind as to sign his hon. name on the top line of the back page of these cards—for the respective humble petitioners who were honored by his presence at 4 a.m. on Sunday morning, May 10th, 1942." FDR charmingly wrote in reply, "Always obedient," then adds his familiar scrawl to the other names, with a proud dash—4 a.m.

At ten, after our walk around the pond where we heard twelve Yellow-throated Vireos singing their raspy song, eleven Least Flycatchers *che-bekking* from a limb, and no Soras or Virginia Rails, where we negotiated narrow boardwalks and got our feet wet, I pulled out cheese sandwiches and hot tea.

"This was epic," Christina said, face spread into a grin.

Christina and I sat on a rock eating, a bit dazed and happy that we had forgotten the war as we listened to and looked at birds since before dawn on May 10, 2013.

LITTLE BLUE

South Tivoli Bay, New York

When Roger Tory Peterson and James Fisher made their birding loop across America in 1953, they visited Avery Island, Louisiana, not far north and west from New Orleans. In their book *Wild America*, Peterson relates the extraordinary story of Edward Avery McIlhenny, a naturalist whose family is known for producing Tabasco sauce. McIlhenny had witnessed the destruction of egrets for the millinery industry and wanted to take action. So on an island the family owned west of New Orleans, he decided to try to protect the Snowy Egret by giving it a perfect place to breed. In 1892 he began to shape his land, building a dam to form a small lake and planting rhododendrons, groves of bamboo, and lotus beds to form an egret Garden of Eden. Over the lake he laced netting to create a large cage. In it he stuck eight young Snowy Egrets, which he had found by scouring the swamps. He fed them shrimp and killifish as they grew tame at his hand. Outside of their cage they approached him without fear; Peterson wrote that they even perched on McIlhenny's horse. After a season or two of breeding, the birds were free to go. They stuck around until the first cold snap, then headed south, despite the lure of hand-served meals. Would they come back?

The first spring six Snowy Egrets returned. From there on the numbers rose dizzyingly: 1905 saw 1,000 pairs; in 1912, 100,000 birds built 22,204 nests. To create all of these messy stick nests they trucked in loads of twigs. It's hard to imagine both so many birds and so many nests, each tree likely bending with the weight of it all. There have never been as many birds as in 1912, perhaps the ecosystem leveling out to something a bit more natural, yet Avery Island remains a bird sanctuary and tourist destination.

I love this story. If I had a patch of land, this would be my dream, to slowly create a place where birds were safe, could breed with ease. It's also great to have a story where protecting the birds is relatively simple: create a lake, add a few buckets of fish and a couple truckloads of twigs. Usually bird preservation is long and complex; the logistics, like stopping heavy moving equipment removing the tops of mountains, or international treaties, even makes protecting some birds impossible. So this story feels solid even though I also know it so easily could have backfired: the lake could have invited in some invasive species, or the twigs might have introduced an unwanted beetle that compromised the birds. We humans are not always so good at engineering the natural world. But Avery was determined and he was lucky. Thanks in part to his vision, Snowy Egrets are still with us.

When, on a hot August day, I spied two white wading birds on the far side of the South Tivoli Bay, my heart did a skip of excitement. The sun shone on the Bay, creating a glistening mosaic. The birds remained tantalizingly out of reach, even looking through my scope. I knew because of the size of the birds that they were not Great Egrets, which are not uncommon in this area. They had to be either Snowy Egrets or juvenile Little Blue Herons, birds I had enjoyed while in Florida, but both species unexpected in the Hudson Valley. The juvenile Little Blue, despite that name, is white, and in size much like a Snowy Egret. The differences between the two species are subtle, except for the feet; the Snowy has great yellow feet that contrast with its black legs. From my distance, all I could make out in the heat shimmer were white wading birds.

As I drove home, I was determined to figure out what was out there in the Bay. This wasn't one of those many so-close-but-yet-so-far unidentified birds I was willing to leave in the field. The next morning I woke at five and decided that to observe the birds at close range, I needed to be on the water. In the predawn light, I drove my sea kayak to the Tivoli launch. Life vest cinched tight, paddles snapped together, I pushed off from the rocky shoreline and took my first strokes; the pink-gold light of the rising sun did not yet give shape to the far shore of the Hudson River. So I paddled in my cocoon of darkness, so familiar with this reach of the river that I knew where rocks lurked beneath the surface, ready to snag my boat, or where I might need to dodge the remains of an old dock.

Wedged between my legs in the kayak was my camera in a dry bag. By now, I had bought myself a decent camera, with a 400 mm lens, smaller than what Peter lugged around, but it would get me closer to the birds. I carried it cradled in my arms on every bird walk and referred to it as Baby, since it weighs as much as a newborn. My first months of photos were pretty pathetic—out of focus, under- or overexposed, and often with only a piece of the bird in the frame. I kept at it, at first in the hope that the photos would help me to confirm what I was seeing. But soon enough, the photos held greater meaning for me, as I saw details in the photos that I did not see while in the field. The photos made me feel like I was keeping close what was best about my life.

Carrying my camera in my kayak was a risk. I had already donated three point-and-shoot cameras to the river, with my most recent river camera falling prey to the wake of a barge. Still, whatever lurked in the South Tivoli Bay I knew was rare enough that I would need my camera to confirm for myself and the world what I saw.

As I pushed south, a train raced by, heading north, blasting its horn as it approached the Tivoli Crossing. I gave a faint wave to the speeding passengers, wondering if anyone was looking out the window. Most passengers nap, stare at computers or phones, chat with a friend. If they saw me I was sure they would feel a pang of jealousy because I couldn't imagine a more perfect place to be than in my kayak on the Hudson River at dawn.

I paddled past cozy and quiet Magdalen Island, then launched into the open river waters before reaching the western border of Cruger Island. An immature Bald Eagle sat rumpled on a branch of a snag, preening, while Ring-billed Gulls with sleek, athletic wings grazed the surface of the water.

Two miles south of Tivoli, I bobbed through fluky waters to carve into the passage between Cruger Island and South Cruger. Cradled by the two wooded islands, a quiet cove with sandy beaches offers protection from the wind and currents of the river. It's a great place to loiter in a boat. Map turtles hoist out to sunbathe, and shorebirds troll for fish or pick along the sandy shoreline. That morning, a few Mallards paddled the waters while a Spotted Sandpiper worked the shoreline, exposed now at low tide. Spatterdock stood tall in the shallow water, its wide green leaves silted brown from the river's ever turbid water. The river smelled like new-mown hay.

I watched the Spotty, industriously bouncing its butt as it made its way along the shoreline. Binoculars to my eyes, I calmed as my thoughts became all Spotty, focused and determined in its erratic pursuit of a meal.

I had learned that birding from a kayak is not the same as birding from land. When I stopped paddling to look at a bird, my boat continued to glide forward. I had to remain alert to where my boat was drifting and if it might hit something or if I might scare off the bird. In my kayak, unable to stop, I am often deprived of the luxury of lingering long with a bird I want to see. In that sandy bay where I was admiring the Spotty, I knew I didn't risk rocks, but caution made me lower my bins to check my course. And what I saw was that my kayak was sliding in slow motion straight toward a white wading bird. Not more than thirty feet from my bow the bird foraged among the spatterdock.

For a moment, the water, the sky, the spatterdock, and that bird: we were one as I floated there under the gentle sun. I existed out of time, somewhat as in a dream, as if I could reach out and touch the bird, but I couldn't move, heavy in my padded kayak seat. I might have even stopped breathing, not wanting to disrupt any of this moment, wanting to have it extend forever. And then with a gasp, I came awake, realizing: *you have to get a picture.*

My hands shook as I fumbled for the camera, pulling it from the dry bag. I snapped a few pictures; once I was sure I had a decent photo, I put the camera down. Yellow-greenish legs. The feet—were they yellow or dark?

I leaned toward identifying it as Little Blue as I hovered, admiring the beautiful bird that clearly did not fear my kayak or me. The bird foraged, moving with a casual precision. From the spatterdock a second bird emerged, and the two stepped about together, wandering over toward the ducks to find a fish or maybe a frog to eat. Neither of them let me see their feet.

I sat comfy in my kayak for what felt like a luxury of time, my heart happily racing. And then the urge to share my find hit me. I pulled my cell phone from its waterproof pouch to find Peter at home redoing his porch. House chores over birding? The first time I heard this, I felt I might have to stage an intervention. But just as my love of birds had changed me, Peter's love had changed him. The result was a person slightly less obsessed with birding.

"I have just had a religious experience here." I had no other word for it, the birds a sort of miracle.

"Why are you calling it a Little Blue?"

Because it is, I thought. Now I had to articulate what had formed more as a hunch.

"What color are the feet?" he asked. "Snowys have bright yellow feet."

I know. "They are wading." I tried not to laugh.

"What else?" Peter asked.

Oh, so many times we had had this conversation!

"Yellow-green legs," I said with half a laugh. "And a grayish bill, no yellow."

"You have to see the feet." Of all things, how ridiculous that on a wading bird you have to see the feet. In this case, they were below the water's surface in a turbid river.

As one of the birds flew over, I reached for my binoculars, dropping my phone. Somehow, it landed on my sprayskirt and not in the river.

"Wait," I called loud enough that Peter might hear. The feet trailing behind the white bird were not bright yellow.

I gathered up my phone. "No yellow feet," I announced, breathless.

"Nice," Peter said. "I'll expect a post about this."

August 2, 1929, Maunsell Crosby wrote, "A cool, clear morning, wind north, but the ground very parched. At the Mill pond I found 10 immature Little Blue Herons, very tame and beautiful—a new species for the County." Since Crosby's first sighting there have only been a few immature Little Blue Herons in Dutchess County. "It's a RARITY," Peter emailed. "First time I have heard of one around here since I've been birding. They do happen, but not often. *Birds of Dutchess County* says 4 sightings in the 60s, 8 in the 70s, 13 in the 80s, and 3 in the 90s. Real good bird(s)." He had done my research for me. He was jubilant on my behalf and irritated I wasn't posting it promptly (that is, within an hour of getting home).

My hesitation at posting to the list surprised me, my confidence wavering even as I looked at the photos that spelled Little Blue Heron. Telling Peter was one thing; telling the birding community was something else, something I had never done before. *Commit to your ID!* I thought. I wrote the post, read it a dozen times, altering a word here or there and finally paring it down—me, the writer—to one bare sentence that I hoped used the proper etiquette. I did know that nothing in my message transmitted the thrill of finding the birds, the lovely day, the texture of the water, how my kayak glided right up to them. Just the facts: two Little Blue Herons found off of Cruger Island. Then a photo—my evidence.

I hit the send button, the news sailing off to an unknown number of fanatic local birders, who would all scrutinize the photo. Hitting the send button was unexpectedly, nauseatingly hard. *You send dozens of emails every day*, I told myself. *What is the matter?* The matter was perhaps that someone would write and say, no, actually, it's not a Little Blue. But beyond being called out for being incompetent, for being a fraud, this *game* of birding was making me tackle something else.

I doubt many think of me as a person without confidence. I'm the one who travels to the South Pole, who paddles the Hudson River, who lives alone in the woods, who voices her opinions, political and otherwise, with conviction. But we all have our weak spots, and my confidence in my birding ability was being pushed; those two Little Blues were asking me to own what I knew, claim my expertise.

I waited a few hours to check my email. No challenging emails appeared in my inbox, only words of praise and excitement: "Nice find!" "Good job!" There I was, having found my first rare-for-Dutchess-County bird. It wasn't a bird so rare that it was going to draw in birders from across the state, but it was going to find a lot of local fans, those intent on their county lists.

For me, finding the Little Blue was a turning point. I had found it because I had pushed far enough into this land, birded a lot, paid attention with greater intensity. And I had done it alone. I knew what I was doing. My celebration was quiet and short lived. I became only more determined to get up the next morning and see what else was out there.

For those who wanted to see the birds, I gave instructions on where to find them, offered to take people out in my kayaks. The next morning I took Christina out to see the birds. I thought she was going to fall out of the kayak in her excitement. Sharing them with her made them more real.

The last time I saw the Little Blues they were daintily grazing through the muck of the bay. As I sat savoring my time with them, I wished for a moment that like Avery, I owned this wide bay, could set up a Little Blue camp and convince the birds to stick around with an easy meal. But I didn't and I couldn't and maybe that was a good thing, just letting the birds be birds. I watched the birds continue on with their free and precarious lives, unaware as they moseyed off into the spatterdock at the ways that our meeting had so touched my life.

So Much to Learn

North Tivoli Bay, New York

It was a day early in June when I called my friend Emily.

"We should sleep in the Bays."

"Yes," she responded without saying hello. Emily has always been good for an adventure or misadventure. We had traveled to Alaska together, paddled and camped the Hudson River, hiked and snowshoed the Catskills. She doesn't care much about birds, but she knows the Bays as well as anyone, all of the crumbling stone or wooden buildings and the stories they hold. Together we had formed a group, Friends of Tivoli Bays, to tend to this thousand plus acres that run from the village of Tivoli south to Bard College, bordering both the North and South Tivoli Bays. Emily led history tours, animating old houses with fun facts and with characters with names like Benthusen and dePeyster. I walked with students into the Bays to show them snapping turtles and beaver paddling about and made sure they noticed a few birds. I loaded friends into canoes to show them the ruins on South Cruger Island, near where I had found that Little Blue

Heron. The Tivoli Bays was where, every spring, Emily and I organized our garbage cleanups, now both on land and by water, every spring removing dozens of bags of junk, stuff, lost treasures.

Now I wanted to sleep in the Bays, to join the birds to watch the moon. What might I learn by sharing the night with the birds?

"I've found a blind that I think is good. It's flat, sort of. And big enough," I told Emily.

Duck blinds punctuate the North Tivoli Bay. Most are simple wooden platforms in various states of disrepair. Hunters are permitted to fix up a blind, but they are not allowed to build new ones. All are gray from age and sun, and one season I watched as a stand crumbled into the water. The stand I wanted to camp on, about eight feet squared, and propped on wooden stilts above high tide by just a foot, had a floor with only one board missing. A slim, jagged piece of plywood might be called a roof.

As I prepared my gear, I realized I couldn't imagine having such a great bird experience without Christina. She dropped everything and was at the wooden dock when we arrived to launch our sea kayaks at 5:00 in the evening. It was high tide as we made our first strokes out in the Bay, the water drowning the spatterdock and pushing up to the edge of the cattails. We meandered toward the stand, poking our way down the wide water alleys of the Bay. There I was, floating with my flock.

We tied up our boats and unloaded our gear. As we spread out our sleeping bags, shoulder to shoulder, in the tight space, joking about who had to sleep in the middle (me), the sun crested over the Catskills, creating pink and orange swirls.

The view from the blind, skimming over a field of grass and cattails, was beautiful. The Bays are intimate, less wind driven than the open Hudson River where barges plow north and south. In the Bays life calms.

Before cozying in with tea, a bag of popcorn, and conversation, Emily reached up to reposition the sagging roof. It fell off in her hands. We laughed nervously as she put it aside.

"Maybe we should fix this blind?" I wouldn't sleep well thinking that I was helping out a hunter, but I'd like to know that this watery tree fort existed for people to sleep in from time to time. "What do you think it would take?"

"Work," Emily said. She is a carpenter who restores historic houses, so she should know.

Soon, the Great Horned Owls began their calls from the forest to the north. First the male, a deeper, heavier hoot. Then the female, higher pitched. There is nothing more heartwarming than an owl song in the night. Except, perhaps the clack of a Virginia Rail, emerging from the reeds.

"Hear that?" I said.

Christina smiled the loony smile of someone possessed by the night sounds, while Emily dug into the popcorn.

"*Strigiformes*," Christina said. "*Gruiformes*."

"*Strigidae. Raillidae.*" I countered. That spring, Christina had enrolled in an ornithology class that I also audited. I had been curious what a scientific perspective might offer to my understanding of birds. Mostly, I gathered intriguing words, like *zugenruhe*, that restless gathering of flocks of birds before they migrate. Or the Latin names for the birds, which Christina and I used mostly to make each other laugh. I couldn't see a cormorant without saying, "Phalla-crack-e-rach-e-day" (*Phalacrocoracidae*).

Sitting on the platform, I remembered when I first kayaked at night on the Hudson River. I rounded the end of Magdalen Island, and a Great Blue Heron left its roost, *craaaaking* its displeasure with me. The sound amplified over the water made me almost topple over in fright. Paddling at night was thrilling and disorienting, the speed and depth of the water a risky unknown. Birding at night felt similar. I sat on the hard wooden platform, alert. Moon-produced shadows played with my imagination; birds drifting on the water became logs or nothing at all.

Though I knew we were safe, cocooned on our wooden island, I still half listened through the night. I imagine this is what the birds do as well, one ear and one eye cocked for that marauding coyote or feral cat. But my half sleep was less about fear and more about a happiness I associated with childhood, the simple thrill of doing what you want to do. It was like those nights under the pine trees in front of my childhood home, the giddiness of sleeping outside with my sister, the darkness of the night a tonic that did not allow for sleep.

I rolled over and the stand shook. I knew Emily and Christina were awake as well.

"Do you think this thing is solid enough to hold us?" I mused, knowing my question was silly.

"I removed an entire section of the roof with my bare hands," Emily said.

A freight train made a long transit, clanking its way south. A barge on the river blew its horn as it pushed north. The noise through the night never stopped. And in the nighttime medley, the sound that stood out was the song of the Marsh Wren reverberating through the reeds.

It's difficult to glimpse one of these spunky birds, but it's not hard to hear them; it's the sound of the Bays. Because it's such an odd, rippling song, many have described it. In reading these descriptions I felt like I was gathering a party of the past birders who informed and shaped my birding experience. There was my poet Alexander Wilson writing that the Marsh Wrens' songs "are bubbles forcing their way through mud or boggy ground when trodden upon." And precious John James Audubon: the "song, if song I may call it, is composed of several quickly repeated notes, resembling the grating of a rusty hinge, and is uttered almost continuously during the fore part of the day." The wren is not a fancy dresser like Audubon, so he describes it as a "homely little bird." And then my godmother, Florence Merriam Bailey: "However much you are prepared for it by other members of the choir [swamp singers] the first outburst of the Marsh Wrens is almost paralyzing. You feel as if you had entered a factory with machines clattering on all sides." Florence! I wanted to say, be more generous in your description of our little brothers of the swamp. But maybe her feelings are influenced by the fact these are not sweet little birds. They are energetic and aggressive, often destroying the eggs of other birds, including those of other Marsh Wrens.

Because I associate the Marsh Wren with the oozy, cattail swampy, rich, secretive North Tivoli Bay, I could not imagine much that would give me more joy. It was as if the Marsh Wrens were singing: be here now; live this life fully.

If the birds did, so should I. And if I did, what would that mean? Nothing special, nothing different. What it would mean was doing exactly what I was doing. It had taken me half a lifetime to find myself sleeping on a platform in the North Tivoli Bay with two friends, to know that sleeping on a platform in the North Tivoli Bay was just where I needed, and where I wanted, to be.

"They never shut up," Emily moaned the next morning.

"Isn't it great?"

We packed up our sleeping bags then slipped our boats into the water. We poked around the Bay, listening to the Swamp Sparrows, the Redwinged Blackbirds. The need for coffee and the hope for a brioche from the Tivoli bakery pushed us all toward the dock, where in a dazed state we loaded our boats onto our cars.

Through the day, I moved as if in water, everything washing over me. In my subdued state, I found myself speaking slowly, walking slowly, thinking slowly. What I thought about was that three years ago my dream had been to learn the birds. Whatever I thought that meant, I never imagined it as rich and wondrous as communing with Great Horned Owls, speaking with rails clattering in the reeds, discovering that Marsh Wrens don't shut up. On this soft, dreamy day I understood that if birds were near, I would never be tired or bored or lonely; if birds were near, I would always have more to learn.

Acknowledgments

To have an editor who edits, line by line, is a rare gift. Michael McGandy has pushed me and believed in my writing since our first book together. I feel lucky to call Cornell University Press my publishing home, and I want to thank all who helped to shape and bring this book into the world: Susan Specter, who has held my hand through this and other books, Deborah A. Oosterhouse, who asked just the right questions so I'd be more precise, and Clare Jones, who is a miracle of efficiency.

Those who read drafts and then more drafts can't be thanked enough: Boyer Rickel had to sharpen his pencil many times; Donna Steiner has for this book and in the course of my writing life read millions of words; Phil Pardi read and offered perfect advice when I needed it. It makes sense that I turned to the poets for help.

Cousins! Thank you, Lisa Redburn, for all of our journeys on beaches or through woods to capture images and birds; your photos and thoughts on photos are always an inspiration to what I do with words. Peg Mitchell and Polly Talen—that you *asked* to hear chapters while sitting on the porch of the cabin on Eagle Nest Island meant the world to me. Your listening, that you did not fall asleep, and those little grunting noises of approval changed the course of this book.

Deb Addis, you have been an inspiration in my bird journey, both showing me the desert and combing my Tivoli Bay woods with me. For this book, you are my perfect reader. Teri Condon, thanks for morning therapy sessions, making me laugh, and sometimes letting me stop to see birds. Emily Majer, thank you for all of our adventures and misadventures in the Bays—let's keep getting lost together. Dinaw Mengestu, you said just the right words of encouragement at just the right time. Thank you for sharing your family—la belle Anne-Emmanuelle Robicquet and the adventure-loving Louis and Gabriel. You all kept me focused on what matters as I finished this book. And thank you to my many amazing, brilliant colleagues at Bard College; your support and good humor are the foundation of all I do.

Playa Summer Lake offered time and space to write in a bird-delicious location in southeastern Oregon. Many words were written in the comfort of my cabin while watching Prairie Falcons rip across the playa. The Ucross Foundation in northeastern Wyoming offered the ideal place to begin thinking about/writing this book.

Thank you to Taza Schaming for her time telling me about her work with Clark's Nutcracker, and Tina Green for that snapshot of finding a rarity. I hope our paths cross in the field some day.

Friends far and near, birder and not-yet birder, thank you! Matilda, Harriet, Amanda, and Tim Arkell, who light up the Ridge in every way (keep the chard deliveries coming); Steve Bauer, who puts up with my lousy IDs and sends me bird poems; Nancy Bissell, who welcomes all of the wandering birds in Tucson; Sonia Caton and Ellinor Michel, my "oldest" friends, from afar you make me feel like my writing matters; Mark DeDea, Lin Fagan, and others from the John Burroughs Natural History Society, who helped to launch me on this path; Ellen Driscoll, our chin wags sustain me, make me laugh, make me want to write; Eli Dueker, your moral compass guides me; Sue Fey, who is my Audubon family; Erik Kiviat, who knows all of the secrets of the Tivoli Bays and shares them so generously; Ryan MacLean, for Troppy chasing, and CBC and May census *weeee*; Jody Melander, for years of adventure on the Cape searching for birds and otters and the meaning of life; Sara Mednick, Jesper, and

Violet and our adventures—on the river and of the heart—you have made my life brighter; Barbara Mansell, Debi Van Zyl, and other local birders remind me Dutchess County is the place to bird; Charlo Maurer, who listens so keenly and always tells me to "do it"; Kate Shaughnessy who is ready for any adventure; Barbara Sproul, who sends me bird books, and who gave me my most treasured Forbush volumes; Feli Thorne, who feeds me in all ways and who gave me my bird ears.

For being there when the Veery sang, thank you Sam Logan and Ali Sickler. To all of my students who have birded the Bays with me, your energy and enthusiasm is at the heart of this book. Gowri Varanashi, Xaver Kandler, Tierney Weymueller, Will Santora, Grace Drennan, and Liza Birnbaum, you and others remind me that learning is energizing, exhausting, and the only way to live. For my students who do not bird with me—where are you?

To Mary McCollum, Gray Glenn, and the Schoenberger family, thank you for your warmth.

Sam Ace, love always.

Alice and Thomas, I could not love your kindness, intelligence, humor, and glorious spirits more. You inspire me and make my life so colorful.

Peter, you were and still are my best teacher. Your enthusiasm for the birds and your generosity in sharing your knowledge is unmatched. Thank you for leading me into a life that I love. Christina, you have added great chapters to my life, both the bird life and the other life-life. Continue to be your irrepressible, wacky, creative, beautiful self.

My sister Becky, writer, historian, runner, remarkable chef, sometimes birder (and great spotter of the brown-muzzled snapchat), amazing reader, endless supporter, thanks for being there through it all. You are my Sweet Baboo.

NOTES, NOTES ON NOTES, AND
FURTHER MUSINGS

A note on common bird names: I follow the American Ornithological Society's system for capitalization of common bird names. For example, Hermit Thrush instead of hermit thrush; Cedar Waxwing, not cedar waxwing. I use lowercase when referring to a group of birds (the owls), or when I'm not referring to a specific bird (the larks). But if I'm looking at a specific Horned Lark, it will be shortened to Lark. If a bird is known commonly by a shorthand, I use that shorthand and capitalize it. For example, Gray Catbird would be Catbird, or American Robin would be the Robin. This means that most often "Eastern," "Northern," or "American" is left off of a name (though never for the American Bittern!). If there is any chance of confusion, I refer to the bird with its full name at first mention in a chapter (for instance Eastern Bluebird) and thereafter use simply Bluebird.

I used an enormous range of sources in researching this book. I know that, despite all attempts at accuracy, there will be errors. Knowing birders, I know I'll hear about them. Please be in touch: susan@susanfoxrogers.com.

I Wish I Knew

My trip to the Amazon was thanks to my former student Gowri Varanashi, who has been a steady bird partner since her time at Bard College. She, along with Paul Rosolie, invited family and friends on this trip through their adventure company, Tamandua Expeditions. Both Paul and Gowri are active in conservation, particularly of the Amazon. And both love snakes.

The book that my student Ali gave me is *The Backyard Birdsong Guide*, by Donald Kroodsma, published as part of a series of audio guides by the Cornell Lab of Ornithology (Chronicle Books, 2008). At the time I had no idea of what a role the Lab of Ornithology at Cornell would play in my birding life—they are bird central. This guide also introduced me to Kroodsma's work, and in particular his *Listening to a Continent Sing* (Princeton University Press, 2016), about bicycling across the country with his son while listening to birdsong, is one that made me dream.

The description of the Veery, "the chiming of bells or the gentle sobs of organs" as well as the "spiral of white gold," comes from a graceful, short essay by Raymond Deck, "Salute to a Brown Bird," found in Roger Tory Peterson's edited collection *The Bird Watcher's Anthology* (New York: Harcourt, Brace, 1957), pp. 93, 91. Deck's essay is one of my favorites from that big book as Deck writes of the heroic migration of the Veery— only to have it die in a most ordinary manner. Peterson describes such an unnecessary death by comparing it this way: "It is as though an intrepid explorer, on his return from far adventures, slipped on the stairs of his own home or killed himself while cleaning his rifle" (p. 90).

Frank Chapman's *Autobiography of a Bird-Lover* (New York: D. Appleton-Century, 1935) is a delightful read and contains the moment of his conversion to the birds on p. 36. My affection for Augustine's *Confessions* emerges from years of teaching this book in Bard College's First Year Seminar Program.

John Burroughs's essay "The Art of Seeing Things" offers guidance on learning to see. Like any art, it requires some talent, and lots of work. The essay appears in his collection, *Leaf and Tendril* (New York: Riverside Press, 1908). Quotes from the essay come from p. 24.

My first inspiration to write about my bird journey came from reading *Fly Fishing through a Midlife Crisis* by Howell Raines (2006). When

I started to bird I might have been going through a midlife crisis, but if a midlife crisis is detected by sudden, drastic life choices (the red sports car, the divorce) then I have had more than one midlife crisis, and none of them at midlife.

Photo of the Veery taken by Peter Schoenberger at Wilson State Park, Ulster County, N.Y.

Snow Bunting

The landscape of this chapter—the Ashokan Reservoir, and the former cement factory along the Hudson River—is one I call home, the Hudson Valley. It's mostly farming country, framed by the Catskill Mountains. As I walk fields or forests, it's hard to believe that Manhattan is only one hundred miles away, but the presence of the city now, as in the past, is vibrant, from the money of weekenders to the trains that lace the region for commuters. Manhattan is built from Hudson Valley bricks and cement, our farms ship fruits and vegetables south, and our water flows into city apartments. This movement of bringing food and water to people is the inverse of what birds do: fly to their food and water. That the Snow Bunting spends its winter in the Hudson Valley means that it finds enough food here to survive the winter, but it's never an abundant bird in this region.

Throughout my research I relied for scientific information on species from Birds of North America Online (birdsna.org; now renamed as Birds of the World, birdsoftheworld.org), which I subscribed to (and it's worth every penny!). Information on the decline of the Bunting I took from Robert Montgomerie and Bruce Lyon, "Snow Bunting (*Plectrophenax nivalis*)," Birds of North America Online, March 4, 2020.

I first heard the story of the Ashokan Reservoir from my student Ali, who wrote about her family losing their home, their village, when the reservoir was completed in 1915. More information on this reservoir, at the time the largest in the world, can be found in Bob Stueding's *The Last of the Handmade Dams: The Story of the Ashokan Reservoir* (New York: Purple Mountain Press, 1989).

The "stuff [that] oozes out of the ground" that Peter referred to in East Kingston I learned from Erik Kiviat, naturalist and founder of Hudsonia, is, in fact, "star jelly." Star jelly is the cyanobacterium *Nostoc commune*.

It is an organism "abundant on highly calcareous sparsely vegetated mineral substrates at the long-abandoned cement quarry complex in East Kingston." When moist it is oozing and green; when it dries out it looks like a crust of algae in shore wrack or "perhaps dried very runny cow manure."

Photo of the Snow Bunting taken by Peter Schoenberger on Hurley Mountain Road, Ulster County, N.Y.

Learning the Birds

Before my bird journey I had never heard of Alexander Wilson, perhaps because John James Audubon, thanks to the national society, looms so large. In Wilson I found a man with range: poet, teacher, and bird man, and someone much humbler than Audubon. Once I realized the rich history of birding I was grateful for Scott Weidensaul's *Of a Feather: A Brief History of American Birding* (New York: Harcourt, 2007), which is not brief, but rather the perfect amount of information on this rich history, and it is beautifully told.

That Wilson's spark bird was the Red-headed Woodpecker is recounted by many, for instance by Edward Howe Forbush in his guide *Birds of Massachusetts and Other New England States* (Norwood, MA: Norwood Press, 1929) but more importantly by George Ord in his *Sketch of the Life of Alexander Wilson, Author of the American Ornithology* (Philadelphia: Harrison Hall, 1828). In it Ord writes: "The writer of this biography has a distinct recollection of a conversation with Wilson on this part of his history [when he first arrived in America], wherein he described his sensations on viewing the first bird that presented itself as he entered the forests of Delaware; it was a red-headed woodpecker, which he shot, and considered the most beautiful bird he had ever beheld" (p. xviii–xxix). Spark stories have a bit of prophetic wisdom to them—we need someone like Wilson to have such a tale. And yet in an April 2016 article titled "Wilson's Redhead" in *Birding* (a magazine published by the American Birding Association) author Rick Wright contests the idea that the Red-headed Woodpecker drew Wilson in. Alexander Wilson's entry on the Red-headed Woodpecker appears in his *American Ornithology; or, The Natural History of the Birds of the United States* (Philadelphia: Bradfield and Inskeep, 1808), 1:142–46, which is also available online through the biodiversitylibrary.org.

For more on Wilson's life, Clark Hunter offers a wonderful biography and has edited Wilson's letters in *The Life and Letters of Alexander Wilson* (Philadelphia: American Philosophical Society, Independence Square, 1983). The quotes by Wilson appear on the following pages: "with a kind of creative enthusiasm," p. 72; "bird of passage," p. 315; "to bale her, and to take my beverage from the Ohio with," p. 326; "two thousand million Passenger Pigeons," p. 98; and "Even poetry, whose heavenly enthusiasm I used to glory in, can hardly ever find me at home, so much has this bewitching amusement engrossed all my senses," p. 212.

For a gripping account of Tim Gallagher's sighting of, and subsequent search for, the Ivory-billed Woodpecker, read his *Grail Bird: The Rediscovery of the Ivory-billed Woodpecker* (New York: Harcourt, Brace, 2005).

Photo of the Red-headed Woodpecker taken by Peter Schoenberger at Esopus Meadows, Ulster County, N.Y.

Bicky

John Burroughs is my local nature writer, and one I have become fond of through proximity. He was a moody man, full of love for the natural world. During his day he was famous, popular with readers from his first book, *Wake Robin* (1871), which does not refer to the bird but rather an early spring flower, most likely a trillium. He came from the Catskills and lived near the Hudson River, but he did travel—to Alaska (with the financier Edward Harriman during his famous 1899 expedition) and to the Everglades (with Thomas Edison and Henry Ford). He was friends with John Muir and was born in the same town as Jay Gould. What I have most enjoyed reading are his journals where the man in all of his complexity emerges. Most of his books are now out of print but available online in various formats.

Burroughs's climb of Slide Mountain is detailed in his essay "The Heart of the Southern Catskills," in *The Writings of John Burroughs*, vol. 9 (Boston: Houghton Mifflin, 1904; repr. Boston: Elibron Classics, 2004). All quotes from Burroughs in this chapter are taken from this essay.

That birds carry the names of those who first "discovered" them is something that I originally found marvelous—imbedded in the names is another layer of history. As my birding life deepened, I started to chafe at these names. I'm not alone in feeling that birds are burdened by the human past. A movement, Bird Names for Birds (with a website of the same name), is seeking to correct this. In 2020 the first of these correctives was made by changing the McCown's Longspur to the Thick-billed Longspur. John McCown was a Confederate general who defended slavery.

The eight birds named for women in the ABA area that I have located include five common names and three Latin names (there may be others, and if so, I'd love to hear about them):

1. Lucy's Warbler. Named for Lucy Hunter Baird (1848–1913). In 1861, James G. Cooper collected (shot) the warbler while working with the Whitney's Geological Survey. He dedicated the bird to the "interesting little daughter of my kind friend, Prof. S.F. Baird." Only thirteen years old at the time, Lucy grew up with snakes and squirrels as pets. She devoted her life to her parents and when her father died set about writing his biography. She never finished; William Dall completed the enormous task in 1915.

2. Virginia's Warbler. Named by Spencer Fullerton Baird for Mary Virginia Anderson (1833–1912). Mary Virginia was married to William Wallace Anderson, assistant surgeon in the U.S. Army. Anderson collected (shot) the warbler in New Mexico in 1860.

3. Blackburnian Warbler. It's not clear if Thomas Pennant named the Blackburnian Warbler for Anna or Ashton Blackburne, or for both of them. I'll include the Blackburnian here and claim it for Anna, making it the only bird that uses the woman's last name. Anna remained in England, while Ashton traveled the United States collecting (shooting) specimens to send to her. So here we have one name related to a woman engaged in natural history. It's too bad it's just such an awful name.

4. Grace's Warbler. Grace Darling Coues was sister to Elliott Coues. He's the one who collected (shot) the warbler at Whipple Pass in Arizona. Coues sent the skin to Baird asking that it be named for his sister, in the hope that "my affection and respect keep pace with my appreciation of true loveliness of character."

5. Anna's Hummingbird. Paolo Botta, a young Italian surgeon, collected (shot) the Anna's Hummingbird during a three-year voyage on *Le Héros*, which included months of cruising the West Coast. He gave the bird to

Victor Masséna, Duke of Rivoli, who then showed it to the naturalist René Primevère Lesson. In gratitude, Lesson named the bird for Victor's wife, Anna Masséna, Duchess of Rivoli. The Rivoli's Hummingbird, previously known as the Magnificent, is named for the duke.

6. White-winged Dove (*Zenaida asiatica*), Mourning Dove (*Zenaida macroura*). Born 1801 in Paris, Zénaïde Bonaparte had the curse or blessing of being a part of the Bonaparte family—her uncle was Napoleon. She married her cousin Charles, a self-taught naturalist for whom the Bonaparte's Gull is named. Charles wrote a "Geographical and Comparative List of the Birds of Europe and North America" and in so doing created the genus Zenaida, which includes our White-winged and Mourning Doves.

7. Blue-throated Hummingbird (now the Blue-throated Mountain-gem) (*Lampornis clemenciae*). Marie Clémence Lesson, born in Paris in 1796, was a well-educated woman who had trained as an artist of the natural world. In 1827 she married René Primevère Lesson, who had traveled the world serving as a surgeon and naturalist aboard ships (among others, he travelled with Dumont D'Urville, who named the Adélie Penguin for his wife, Adèle). Lesson first described the Blue-throated Hummingbird, and in French named it the L'oiseau-mouche de Clémence.

8. Rose-throated Becard (*Pachyramphus aglaiae*). Aglaé Brelay, born at an unknown date in France, busied herself helping her husband, who had a vast collection of skins. Little else is known of him. She caught the eye of Baron Frédéric de Lafresnaye, a foremost ornithologist, who named the bird for her. That he did so was an exception to his usual practice as he wrote, "We are far from approving of the habit of giving new birds the names of women, who are often strangers to the love of ornithology."

It strikes me that when a bird has a common name of a woman her first name is used: Lucy's Warbler, Anna's Hummingbird, or Grace's Warbler. To truly celebrate these women, it should be Baird's Warbler, the Masséna, Duchess of Rivoli's Hummingbird, and Coues's Warbler. I understand that using Baird might confuse young Lucy with her famous father, or that the Duchess's name might be a mouthful for a bird as small as a hummingbird, and Coues could be confused with her famous brother Elliot. So the solution here is not obvious (then again, getting the vote in 1920 wasn't obvious either).

This information on bird names and the quotes used here are taken from Barbara and Richard Mearns, *Audubon to Xantus: The Lives of Those Commemorated in North American Bird Names* (London: Academic Press, 1992).

Information on the Bicknell's Thrush comes from Jason M. Townsend, Kent P. McFarland, Christopher C. Rimmer, Walter G. Ellison, and James E. Goetz, "Bicknell's Thrush (*Catharus bicknelli*)," Birds of North America Online, August 13, 2015.

George J. Wallace's *My World of Birds: Memoirs of an Ornithologist* (Philadelphia: Dorrance, 1979), offers an insightful narrative of early ornithological work on the Bicknell's Thrush. The quotes I use, "a constant flow of thrush like musical notes" and "perched on the rim of the nest and sang . . .," come from pp. 54 and 52.

Since a Bicknell's Thrush is tough to see, and harder to photograph, the photo here is of the Blackburnian Warbler, seen at the end of my hike. Photo taken by Peter Schoenberger in Ulster County, N.Y.

Methinks

The Grasshopper Sparrow that Peter so mourned in his early texts to me— do not fear. It would take ten years before the Grasshoppers returned to that odd but charmed dump, but return they did. In May 2020, Peter sent me a text announcing: "Ten years on Grasshopper returns!"

In *Hope, Human and Wild: True Stories of Living Lightly on the Earth* (New York: Little Brown, 1995) environmental writer Bill McKibben writes about flying over this northern region of Maine. From the air the dirt logging roads appear framed by trees. Beyond a quarter mile buffer spread "a plain nearly devoid of life." With a "few little clumps of trees and then vast fields" (p. 40), it could have been Kansas.

For information on habitat destruction, I read Jocelyn Zuckerman, "Plowed Under," at *Prospect.org*, February 2013, http://prospect.org/article/plowed-under.

Information on the Nelson's Sparrow comes from W. Gregory Shriver, Thomas P. Hodgman, and Alan R. Hanson, "Nelson's Sparrow (*Ammodramus nelsoni*)," Birds of North America Online, 2011. Though they list the bird as one of "least concern," it is a bird that has never been abundant. In Edward Howe Forbush's 1929 *Birds of Massachusetts and Other New England States* (Norwood, MA: Norwood Press) he writes of the "Economic Status" of the bird: "Probably harmless but too rare to be of any importance in New England" (3:65).

Henry David Thoreau's description of the Bobolink I found in Florence Merriam Bailey, *Birds through an Opera Glass* (published under her maiden name, Florence A. Merriam, New York: Chautauqua Press, 1889), p. 32. I was not able to find this quote among Thoreau's own writings.

Quotes from Henry David Thoreau come from his essay "Ktaadn," which is part one of *The Maine Woods* (New York: Library of America, 1985, originally published in 1864), pp. 593–655.

Wilson's mourning of his childless status appears in Clark Hunter's biography, *The Life and Letters of Alexander Wilson* (Philadelphia: American Philosophical Society, Independence Square, 1983), p. 67.

Photo of the Nelson's Sparrow taken by Peter Schoenberger, at Southlands, Dutchess County, N.Y.

Florence

In writing my story, I wanted to include more women in the conversation on birds and birding, whether from the past or from the present. It was in Deborah Strom's sadly out-of-print anthology, *Birdwatching with American Women*, that I found Florence Merriam Bailey (1863–1948) and many others, but it was Bailey whom I was most drawn to. Bailey wrote eight books and guidebooks and published dozens of articles. I most enjoyed *Birds through an Opera-Glass* (New York: Chautauqua Press, 1889), considered the first field guide, and *A-Birding on a Bronco* (Boston: Houghton Mifflin, 1896), in which she details that spring and summer of 1894. Her guide, *Handbook of Birds of the Western United States* (1902), is illustrated by Louis Agassiz Fuertes and weighs in at over 550 pages. It is in this guide that I found her description of the call of the Clark's Nutcracker (p. 283).

The California Woodpecker referred to in Bailey's story in *A-Birding on a Bronco* is named in "Ridgway's Manual" *Melanerpes formicivorus bairdi*. This then must be a subspecies of Acorn Woodpecker (*Melanerpes formicivorus*) that is no longer listed in modern guidebooks. The baby woodpecker she names Bairdi is obviously taken from the subspecies Latin name, which recognizes ornithologist Spencer Fullerton Baird.

For a detailed and lively biography on Bailey, read Harriet Kofalk, *No Woman Tenderfoot: Florence Merriam Bailey, Pioneer Naturalist* (College Station: Texas A&M University Press, 1989). Florence's letter to her brother asking about what he wanted "done with the skeleton [he] left on the kitchen stove" appears on p. 18, and the quote about Burroughs's wolfishness with the Smith students on p. 37.

There are now many more resources for women in birding. Among them are the Feminist Bird Club as well as the World Girl Birders Facebook group. Both are great resources, and it was on the FB group that readers pointed me to a few guidebooks authored, coauthored, illustrated, or containing photographs by women from around the world. That women are not as present as they might be is something written about and discussed in the bird world, including in Purbita Saha's article, "When Women Run the Bird World," in *Audubon*, available online: https://www.audubon.org/news/when-women-run-bird-world#.

Information on the number of birds killed for the millinery industry appeared in *The Auk*, July 1888, pp. 334–35. Robert Ridgway in his book *The Humming Birds* (Smithsonian Institution, 1892) also mentions the bird and particularly the hummingbird slaughter on p. 256.

The Clark's Nutcracker that I found so special on that hike in the Bighorn Mountains is, in fact, not such a remote and elusive bird. A few weeks later I was a little dismayed when, while camped in Yellowstone Park, a Clark's Nutcracker perched above my picnic table eyeing my canned ravioli (which was nearly inedible in any case). Nothing but your ordinary camp robber. As Schaming pointed out: they are a corvid, very adaptable, even to canned ravioli.

Since I interviewed Schaming in 2015, she completed her PhD on the Clark's Nutcracker, receiving her degree in Natural Resources from Cornell University in 2016, gave birth to her first child, published over a dozen articles in both scientific and popular publications, and launched the Nutcracker Ecosystem Project (thenutcrackerecosystemproject.com). She also admits that she has "fallen in love with the Nutcracker since beginning to study them, even though that wasn't why [she] started the research."

I took the photo of the Clark's Nutcracker when it came to steal my dinner in Yellowstone National Park.

Twitching

Since this Cove Island twitch in 2010 I have embarked on a few other, mostly unsuccessful, twitching adventures—Red-billed Tropicbird off of Maine (twice) and Eared Quetzel in the Chiricahuas of Arizona—both with no luck. And I have to say that searching for a special bird that another has found can be fun, but it's not my preferred birding. Still, I love the energy of those who chase (mostly my younger bird friends), and though I can't keep a list I admire those who do.

Mark Cocker's *Birders: Tales of a Tribe* (New York: Grove Press, 2001) remains one of my favorite bird books for his humor and his ability to string together more adjectives than should go together. The ones I quote come from pp. 50 and 133.

My interview with Tina Green was conducted on June 20, 2015, so several years after her discovery of the Fork-tailed Flycatcher. Since then, her rare bird findings in Connecticut have continued with a Gull-billed Tern, a Painted Bunting, and a Ruff in 2016, a Sabine's Gull in 2017, and a Franklin's Gull in 2018.

Photo of the Fork-tailed Flycatcher seen in Connecticut taken by Peter Schoenberger.

Christmas Bird Count

Christmas Bird Count has become my favorite holiday, so much so that I now participate in three local counts, one where my sector includes my yard (it is great to wake up and start listening for owls from the warmth of my porch).

There are many narratives of CBC, but George Plimpton's "Tsi-lick Goes the Henslow's: And Other Gleanings of an Amateur among Professionals on the Christmas Bird Count," which appeared in *Audubon*, November–December 1973, remains one of my favorites.

Glenn Proudfoot continues his Saw-whet Owl research to this day, publishing many articles. The one I mention here he coauthored with Sean R. Beckett. "Large-scale Movement and Migration of Northern Saw-Whet Owls in Eastern North America," *The Wilson Journal of Ornithology* 123, no. 3 (2001): 521–35.

Theodore Roosevelt compiled his bird list that contained the Saw-whet Owl in 1908, supposedly from memory. The list was published in *Bird-Lore* 12, no. 2 (March–April 1910).

Photo of the Saw-whet Owl seen on Amherst Island, Canada, taken by Peter Schoenberger.

Don't Move

I referred to a range of guides as I learned the birds. The guide I use in the field is David Allen Sibley, *The Sibley Guide to Birds*, 2nd ed. (Boston: Houghton Mifflin, 2014), referred to by all simply as *Sibley's*. The early guides of Ridgway and Bent and Bailey often helped me to think about how people wrote about and described birds. But for pleasure I turned most often to Edward Howe Forbush and his three-volume *Birds of Massachusetts and Other New England States* (Norwood, MA: Norwood Press, 1929). It is the most energizing, the most enjoyable to just sit and read because his glee is palpable. His guide moves tidily through each species, giving the expected information on size and colors, songs and diet. Then he delivers up the real information, telling stories as if we were in the field with him, in a section he labels "Habits and Haunts." Forbush offers tales from his own experiences with the birds as well as those sent to him by others in the field. Forbush sheds his scientific lens, becoming a "footloose reporter" on the birds. If Edward Howe Forbush's prose is occasionally overblown, this "results from a genuine ecstasy in the man, rather than from lack of discipline." These words come from the great stylist E. B. White, who cataloged few greater sins than overblown prose. Yet he loved Forbush, and in a rich essay titled "Mr. Forbush's Friends" (*The New Yorker*, February 26, 1966; reprinted in *Essays of E.B. White* [New York: Harper & Row, 1977], p. 262) introduced many, including me, to Forbush. White turns to Forbush, when he is "out of joint, from bad weather or a poor run of thoughts." So I too often turned to Forbush for a better mood and great insight into the birds.

There are many biographies of Audubon, all more rigorous and detailed than that written by John Burroughs. But I chose to rely on the Burroughs biography (*John James Audubon*, Boston: Small, Maynard, 1902) in writing my story because of my familiarity with Burroughs and because he has a more informal, more biased view of this important early bird enthusiast (quotes from this biography appear on pp. 125, 7, and 89). Burroughs is at once full of admiration for the man and deeply critical. For accuracy of information biographies of Audubon beyond Burroughs should be consulted.

Keats's grandson, who was so critical of Audubon, is quoted in Lawrence M. Crutcher, *George Keats of Kentucky* (Lexington: University Press of Kentucky, 2012), p. 74.

Audubon's time in Kentucky, mostly before his bird-devoted life, appears in L. Keating, "Picaresque Encounters," in *Audubon: The Kentucky Years* (Lexington: University Press of Kentucky, 1976). The quote from John Keats comes from p. 78.

Audubon's essay "Myself," which he wrote to get his life story right, primarily for his children, is a great read as he is self-aware and at times charmingly self-critical. His moment of turning to his talents I read in *John James Audubon: Writings and Drawings* (New York: Library of America, 1985), "Myself," p. 792. His confession of his fondness for fancy dress appears on p. 783.

Both Arthur Cleveland Bent, in *Life Histories of Familiar North American Birds: Birds* of Prey (Part 2), United States National Museum Bulletin 170 (Washington, DC: Smithsonian Institution, 1938) and Edward Howe Forbush in *Birds of Massachusetts*, vol. 2, *Land Birds from Bob-Whites to Grackles*, offer brilliant descriptions of the Long-eared Owl, its behavior, and vocalizations that I quote from here.

Photo of our Long-eared Owl found near Bryn Athyn, Pennsylvania, taken by Peter Schoenberger.

No Other Everglades

Two years after our trip to the Everglades, a King Rail appeared in the Hudson Valley at a swamp north of the town of Saugerties called the Great Vly where Peter and his wife Amy have purchased a piece of land. Peter sent me an email: "Now that I've seen this King, I think our Florida bird was a Clapper." That's how long he mulled over the identity of these Florida birds.

Roger Tory Peterson and James Fisher mention Bradley in *Wild America* (Boston: Houghton Mifflin, 1955), on p. 127.

I found the wonderful description of Cape Sable augerdent in Red Smith, "Birdman of the Glades," *New York Times*, December 4, 1977, p. 3.

The story of Guy Bradley reads like a novel and is written by Stuart B. McIver in *Death in the Everglades: The Murder of Guy Bradley, America's First Martyr to Environmentalism* (Gainesville: University Press of Florida, 2003). Quotes from the book appear on pp. 20, 38 and 141. The quote from William Earl Dodge Scott, who was an ornithologist who traveled to the Everglades in 1886, reads in full: "A perfect cloud of birds were always to be seen hovering over islands in the spring and early summer months, and conspicuous among them were brown pelicans, man-o'-war birds, reddish egrets, Florida cormorants, Louisiana herons, American egrets, snowy herons, little blue herons, great blue herons, and both kinds of night herons. . . . It was truly a wonderful sight, and I have never seen so many thousands of birds together at any single point." A little over a decade later, he returns to the same spot to find "only a few cormorants. . . . I found no other large birds breeding, absolutely not a single pair of herons of any kind." Scott described trees full of nests, "some of which still contained eggs, and hundreds of broken eggs strewed the ground everywhere. Fish crows and both kinds of buzzards were present in great numbers and were rapidly destroying the remaining eggs. I found a huge pile of dead, half-decayed birds, lying on the ground. . . . I do not know of a more horrible and brutal exhibition of wanton destruction than that which I witnessed here." And I do not know of a more brutal description.

Marjory Stoneman Douglas's name has been, in recent years, connected more to a high school in Florida that was the site of a mass shooting. But in her day, it was her work for the environment that made Douglas famous. She published *The Everglades: A River of Grass* (New York: Rinehart, 1947) as part of the Rivers of America series, edited by Hervey Allen and Carl Carmer. "There are no other Everglades in the world" is the first line in Douglas's book. Douglas championed women's role in environmental work, and the quote about housekeeping comes from an obituary, written by Margaria Fichter, "Pioneering Environmentalist Marjory Stoneman Douglas Dies at 108" (*The Miami Herald*, May 14, 1998).

The epigram from Martial comes from book 13, number 71, "The Flamingo," of his epigrams, available online: https://www.tertullian.org/fathers/martial_epigrams_book13.htm.

Photo of the coy Brown Pelican seen in Flamingo, Florida, taken by Peter Schoenberger.

Little Brother Henslow

For several weekends in the summer of 2011, Peter and I had driven to birding spots two hours from our homes. We visited Bashakill, a marshy area in Sullivan County where we kayaked out to find Least Bittern doing splits in the cattails. We ventured south to what is known as the Black Dirt region hard on the New Jersey border. The dark soil is excellent for growing onions—acres of onions—and on one particularly wet weekend the place smelled like the inside of the refrigerator when I've forgotten the onions in the bottom drawer. Onions or not, shorebirds like the open fields. Another weekend we pushed north to the Montezuma National Wildlife Refuge for another tour of shorebirds. I mention all of this to point out the many rich bird habitats all in New York State and all within a short driving distance of home.

We had started this trip north toward the Henslow's at the Partridge Run Game Management Area near Rensselaerville, New York. It is a dark, wooded area laced with narrow dirt roads that used to be an apple orchard, now abandoned. Partridge Run is never particularly birdy—but we had heard the mew of the Yellow-bellied Sapsucker, the melodious Rose-breasted Grosbeak, the high whistle *swee* of the flashy Blackburnian Warbler, the insistent Chestnut-sided Warbler, calling *wit wee-chew*. Through the sun-sprinkled summer-green leaves, a galumpfing creature took shape. I first imagined dog, then realized it was too big, and dark. A first-year bear crossed our paths, making it, after the Henslow's, the best bird of the trip.

The wooden boat I thought I would never build—I did. I finished building my Annapolis Wherry after five years of sanding and more sanding.

I originally read the *Doonesbury* cartoon in the *International Herald Tribune* in 1982 and found it in a pile of newspapers waiting to be burned in our family home thirty-five years later.

Olive Thorne Miller's enthusiasm for the work of Michelet appears on p. 204 of her *Bird Ways* (Boston: Houghton Mifflin, 1885), a book that takes on the plume trade (p. 204) and details her purchase of her pet birds (p. 27), as well as her intimate descriptions of birds, like the Wood Thrush (p. 22). Her method of observing the birds I read about in an article by Mrs. M. Burton Williamson, "Some American Women in Science," in *The Chautauquan: A Monthly Magazine* 23, no. 4 (January 1899), p. 363.

Describing bird song is a challenging affair. That Olive Thorne Miller describes the Western Wood-Pewee as a "a droll and dismal affair" in *A Bird Lover in the West* (Boston: Houghton Mifflin, 1900), p. 18, says, in my mind, more about her and less about the bird's song.

Edward Howe Forbush, in his *Birds of Massachusetts and Other New England States* (Norwood, MA: Norwood Press, 1929), offers glee-inducing descriptions of bird song, including the "mournful but mellow, rolling whistle like that of autumn wind, 'wh-e-e-e-e-e-e-e-e-e-e-o-o-o-o-o-o-o-o-o" of the Upland Sandpiper (1:448) and the Henslow's habit of singing all night long. Those who belittle the Henslow's song include C. M. Jones, reporting "Henslow's Sparrow: Nesting in Northern Conn" in the *Ornithologist and Oölogist* 6 (1881), pp. 17–18.

That Audubon named the Henslow for Henslow is written in his *Birds of America*, which is available online: https://www.audubon.org/birds-of-america/henslows-bunting.

Photo of the Henslow's Sparrow seen outside of Ames, New York, taken by Peter Schoenberger.

Interlude: The Other Leopold

Jared Kirtland was a first-rate naturalist, known to be cheerful, unpretentious, and with a great sense of humor. And he was industrious, producing over thirty new varieties of cherries by hybridization. He was, therefore, known as the Cherry King. Which means that the Kirtland's Warbler could be named the Cherry King Warbler.

The Kirtland's Warbler was removed from the endangered species list in 2019.

Leopold's article on the Kirtland Warbler, "The Kirtland's Warbler in Its Summer Home," was published in *The Auk* 41, no. 1 (January 1924), pp. 44–58. All of the quotes about the young men's trip to find the Kirtland's nest come from this article.

James D. Watson's memoir, *Father to Son: Truth, Reason and Decency* (Cold Spring, NY: Cold Spring Harbor Laboratory Press, 2014) offered a unique perspective on Leopold.

I found Leopold's memoir, *Life Plus 99 Years* (with an introduction by Erle Stanley Gardner; Westport, CT: Greenwood Press, 1957), to be a fascinating read. He begins his story after the infamous murder, detailing his arrest and time in prison, so his time looking for the Kirtland's Warbler is not part of his life story. All quotes come from this narrative.

Photo of the Kirtland's Warbler taken by Christina Baal during the Biggest Week in American Birding, Oregon, Ohio.

Good Bird

Deb and I continue to bird with the same intensity whenever we see each other. In 2020 during COVID lockdown she and Larry found a pair of Eared Quetzels in the Chiricahua Mountains in Arizona. I drove out to Arizona that fall in my camper to try and see these marvels—with no luck.

Roger Tory Peterson's descriptions of the owls in Madera Canyon appear in Peterson and James Fisher, *Wild America* (Boston: Houghton Mifflin, 1955), pp. 229–31.

Photos of a Rufous-capped Warbler or of a Black-chinned Sparrow are as hard to come by as the birds themselves. So the photo here is of a Sandhill Crane, which Deb and I admired by the thousand at Whitewater Draw. A good bird. Photo taken by Peter Schoenberger at the Shawangunk Grasslands, Ulster County, N.Y.

Guided

Alaska has always been and remains the holy grail for those working on their bird lists. When Roger Tory Peterson and James Fisher toured the

United States in their Big Year in 1953, Alaska delivered: "The crucial bird—Number 488—came shortly after we reached Anchorage. Flying overhead near the airport was a short-billed gull" (Roger Tory Peterson and James Fisher, *Wild America* [Boston: Houghton Mifflin, 1955], p. 368). At the end of 1953, Peterson had 572 species, not including 65 seen in Mexico. By the time Kaufman arrived, twenty years later, the numbers in a year list were much higher: Kaufman ended his year with 671 species. He had also received advice about Alaska, from Ted Parker: "You can hit 650 easily if you get to Alaska" (Kenn Kaufman, *Kingbird Highway* [Boston: Houghton Mifflin, 2006], p. 55).

Big Years are getting bigger in geographic scope as well, thanks largely to the Internet connecting birders to local experts who can show them new and special birds in record time. In 2015 Noah Strycker completed a global Big Year, flying across the world and connecting with birders worldwide to find an astonishing 6,042 species.

Our guide in Gambell described identifying the Willow Warbler and the Siberian Chiffchaff (both in the genus *Phylloscopus*) as an old-world (or European) ID challenge similar to identifying *Empidonax* flycatchers in this country. The empids kept Deb and me busy in Arizona.

Months after our trip to Alaska, Peter and I received an email from our guide. After much deliberation, the powers that be determined that the bird *was* a Siberian Chiffchaff, a North American first. And it was one of Peter's "crappy" photos that had helped to make the ID.

The photo is not one of Peter's crappy photos of the Siberian Chiffchaff, but rather a bird we often saw in flight in Gambell, and then saw roosting later in our Alaskan trip in the Pribilofs. The Tufted Puffin is one of the most elegant, diva birds we saw. Photo taken by Peter Schoenberger.

Chiuit

Descriptions of the song of the Bristle-thigh Curlew and information on the bird's migration and nesting all come from Jeffrey S. Marks, T. Lee Tibbitts, Robert E. Gill, and Brian J. Mccaffery, "Bristle-thighed Curlew (*Numenius tahitiensis*)," Birds of North America Online, January 1, 2002.

Photo of the Bristle-thighed Curlew in flight, seen outside of Nome, taken by Peter Schoenberger.

A Perfect Fall Day

Photo of the Red Phalarope, found at Greig Farm, taken by Peter Schoenberger.

Surviving the Winter

For those worrying about the Snow Goose and its fate, it did make it through the winter, and come May, it entered into a threesome with two Canada Geese. Through early summer, five goslings, all fluff and vibrant energy, trailed the three adults who watched over them with a don't-mess-with-me determination. Every time I saw them I laughed at the improbable family that they were. When on a late spring walk I showed them to Christina she set about tracking their lives through the summer, gave them a name (the Snooks), and spoke of them as if they were family.

There are also several guidebooks illustrated by women. These are the ones that I have found: *Aves da Demétria*, illustrated by Lorena Patrício Silva and Gersony Jovchelevich (2017); *Birds of the Top End of Australia*, illustrated by Denise Lawungkurr Goodfellow (2005); *A Guide to the Birds of Mexico and Northern Central America*, illustrated by Sophie Webb (1995). If readers know of others I would love to hear of them.

There are, of course, guidebooks written by women and illustrated with photographs taken by either the author or a photographer.

Rachel Carson's quote comes from her work *The Sense of Wonder* (New York: Harper Collins, 1965), p. 87. And Aldo Leopold's words from his essay "Marshland Elegy" in *Sand County Almanac* (New York: Oxford University Press, 1949), p. 101. Carson and Leopold, though they are not bird people, are both central to my understanding of, and love for, the natural world.

Florence Merriam Bailey's thoughts on the Turkey Vulture appear in *A-Birding on a Bronco* (Boston: Houghton Mifflin, 1896), p. 83. And Buffon's mean-spirited vulture words are quoted in Alexander Wilson, *American Ornithology; or, The Natural History of the Birds of the United States* (Philadelphia: Bradfield and Inskeep, 1814), 9:101. Wilson's entry on the Canvasback appears in volume 8, pp. 103–9. All of Wilson is available

online through the Biodiversity Heritage Library (https://www.biodiver
sitylibrary.org).

William Faulkner's Turkey Vulture-ness is from an interview with Jean
Stein, "Interview: William Faulkner: The Art of Fiction," published in the
Paris Review, Spring 1956, no. 12.

George Ord, who so admired Alexander Wilson and considered he had
great talents for description of birds, was, however, dubious about his
poetry. "Whether or not he ever attained to positive excellence in poetry,
may be a subject of dispute," he writes in *Sketch of the Life of Alexander Wilson, Author of the American Ornithology* (Philadelphia: Harrison
Hall, 1828). It is clear what side of the dispute Ord would take. No. Wilson's heavy-handed Canvasback poem can be found in Frank L. Burns,
"Alexander Wilson: His Early Life and Writings," *The Wilson Bulletin* 22,
no. 2 (1910), p. 88.

Photo of the Snow Goose taken by Peter Schoenberger in Ulster
County, N.Y.

#1 Birder

eBird is a part of every birder's life these days. It helps us to find out where
to bird when we travel, what birds are in a particular area, and who is
birding where. Though I list the birds I see less on eBird than I did when
I first started birding, I still get email alerts every day about birds seen, in
my county, in New York State, and rare birds in the country. I'll stare at
the lists and muse over the Northern Jacana seen in Arizona or the Ruff
in New Jersey. It's a way to vicariously bird travel. Meanwhile, Peter and
Mark (though he is now a Papa and birds less often) still jockey for number one status every year over in Ulster County.

Information on eBird being used in bird conservation has many
sources. I used Jim Robbins, "Paying Farmers to Welcome Birds," *New
York Times*, April 14, 2014.

The *Birds of Dutchess County* is a guide that is regularly updated, most
recently in 2019 by Barbara Butler and Stan DeOrsey. It is available on
the Waterman Bird Club website.

The description of the Prairie Warbler comes from Roger Tory Peterson
and James Fisher, *Wild America* (Boston: Houghton Mifflin, 1955), p. 74.

I quote from Maunsell Crosby's journals, which are available in the Roosevelt Archives. Crosby, Maunsell S. Grasmere, Rhinebeck, N.Y. Roosevelt Archives. FDR: Hudson Valley and Dutchess County Manuscripts. Crosby Maunsell: bird record notebooks: "A Year book of Birdlife at Rhinebeck and Dutchess Co, NY," 1909–1916.

The first chapter of Florence Page Jaques's *Birds across the Sky* (New York: Harper and Brothers, 1942) contains her story of birding with the A team in Dutchess County. All quotes come from this chapter.

Griscom-isms are offered in William E. Davis Jr.'s *Dean of the Birdwatchers: A Biography of Ludlow Griscom* (Washington, DC: Smithsonian Institution Press, 1994), p. 115.

I took the photo of the Prairie Warbler during a May Census near Pine Plains.

Rusty Blackbird

When asked what my favorite bird is, I reply without hesitation: the Rusty Blackbird. Surely some are disappointed I don't have a more glamorous or colorful bird that I dote on. But I do love it, and every time I tell someone about my affection for the Rusty, I feel as if I'm introducing people to this special, endangered black bird.

Robert Penn Warren's poem "Redwing Blackbird" appears in *The Collected Poems of Robert Penn Warren* (1998), but where I read it was in the lovely anthology *Bright Wings: An Illustrated Anthology of Poems about Birds*, edited by Billy Collins (New York: Columbia University Press, 2009), p. 225. The book contains illustrations by David Allen Sibley.

John Burroughs writes about the song of the Rusty in "Wild Life about My Cabin" from *Far and Near* (Boston: Houghton Mifflin, 1904). Where I first read it was in the collection *The Birds of John Burroughs*, edited and with an introduction by Jack Kligerman (Woodstock, NY: Overlook Press, 1988), p. 137. Burroughs refers to the Rusty in his essay as the Rusty Grackle, so it is possible he is not referring to the Rusty Blackbird— but the description to my mind holds. Other descriptions of the Rusty's song come from Dr. Charles W. Townsend (1920), as quoted in A. C. Bent, *Life Histories of Familiar North American Birds: Weaver Finches,*

Blackbirds, Orioles and Tanagers, U.S. National Museum Bulletin 211 (Washington, DC: Smithsonian Institution, 1958), as well as from Edward Howe Forbush, *Birds of Massachusetts and Other New England States*, vol. 3, *Land Birds from Sparrows to Thrushes* (Norwood, MA: Norwood Press, 1929).

Information on the Rusty Blackbird can be found on the Rusty Blackbird Conservation Group web page (rustyblackbird.org). I also turned to several articles, including Russell Greenberg and Steven M. Matsuoka, "Rusty Blackbird: Mysteries of a Species in Decline," *Condor* 112, no. 4 (November 2010), 770–77.

Theodore Roosevelt's letter to Chapman can be found in Frank Chapman, *Autobiography of a Bird-Lover* (New York: D. Appleton-Century, 1935), p. 181.

Rusty Blackbird photo taken by Peter Schoenberger at The Great Vly, Ulster County, N.Y.

Dawn Chorus

Since that first May 10 Thompson Pond outing I've gone out to listen to the dawn chorus every year, bringing students with me, friends, or going alone. There has never been another meeting of the May 10th-ists, but this dawn chorus day, along with CBC, Big Sit, and May Census, has become a part of my most-loved birding traditions.

James L. Whitehead's article, "A President Goes Birding," published in *The Conservationist*, May–June 1977 (pp. 20–23) contains wonderful photographs of the president with his birding friends.

Geoffrey C. Ward, in *Closest Companion: The Unknown Story of the Intimate Friendship between Franklin Roosevelt and Margaret Suckley* (New York: Simon & Schuster, 1995), offers letters and journal entries that share Suckley's perspective on the dawn chorus outing (pp. 154, 158).

William E. Davis Jr. offers the story of Ludlow Griscom in *Dean of the Birdwatchers: A Biography of Ludlow Griscom* (Washington, DC: Smithsonian Institution Press, 1994). Quotes from the book come from p. 129.

Visiting the Roosevelt Library to read the president's bird journal was one of my great research moments. I found the material I quoted in Roosevelt, Franklin D: Family, Business and Personal Papers, Bird Diary, Box 8.

It was the hooting of the Barred Owl that started off our Dawn Chorus day. I took this photo of the owl on the Hopeland trails in Staatsburg, New York.

Little Blue

Since finding that Little Blue in 2012, several Little Blues have arrived in Dutchess county and been seen by a flock of gleeful birders. It's hard to know if this is due to more birds visiting or to more birders scouring the fields and ponds for birds. Perhaps a combination of both.

Quotes from Maunsell S. Crosby come from his journals, located in the Roosevelt Archives. FDR: Hudson Valley and Dutchess County Manuscripts. Crosby Maunsell: bird record notebooks: "A Year book of Birdlife at Rhinebeck and Dutchess Co, NY," 1909–1916.

Photo of the Little Blue Heron I took on the Hudson River south of Cruger Island.

So Much to Learn

The Marsh Wren is a bird that keeps birders busy with descriptions of their vocalizations. Those that I offer in this chapter come from John James Audubon, *Birds of America*, plate 100, available online at https://www.audubon.org/birds-of-america/marsh-wren, and Florence Merriam Bailey, *Birds of Village and Field: A Bird Book for Beginners* (Boston: Houghton Mifflin, 1898), p. 203. The description from Wilson appears in Edward Howe Forbush, *Birds of Massachusetts and Other New England States*, vol. 3, *Land Birds from Sparrows to Thrushes* (Norwood, MA: Norwood Press, 1929), pp. 350–52. Here are a few other descriptions of the Marsh Wren from Forbush that delight me:

P. L. Hatch: Indistinct rasping or grating sound like that produced by "a sliver on a fence rail vibrating in the wind." (This description makes my teeth hurt.)

C. W. Townsend: "The song begins with a scrape like the tuning of a violin followed by a trill which bubbles, gurgles, or rattles, depending no doubt on the skill or mood of the performer; at times liquid and musical, at other times rattling and harsh, but always vigorous. It ends abruptly but is generally followed by a short musical whistle or a trill as if the Wren were drawing in its breath after its efforts."

A. Allen: "From a tangled mass of brown cat-tails comes a peculiar grinding sound as though someone were gritting his teeth. This is followed by a clicking noise much like an old-fashioned sewing machine, and then out of the flags bursts a little brown ball. Floating upward like a tuft of cotton, it breaks into vivacious music and then drops back into hiding to continue its scolding."

Pete Dunne: "The sound of a cassette fast forwarding." (Will young birders know what this sounds like?)

I took the photo of the Marsh Wren in the Tivoli Bays.